Dear
Mr Darwin

Dear Mr Darwin

LETTERS ON THE EVOLUTION OF LIFE AND HUMAN NATURE

Gabriel Dover

UNIVERSITY OF CALIFORNIA PRESS
Berkeley   Los Angeles

University of California Press
Berkeley and Los Angeles, California

Published by arrangement with
Weidenfeld & Nicolson

Library of Congress Cataloging-in-Publication Data

Dover, G. A. (Gabriel A.)
    Dear Mr Darwin: letters on the evolution of life and human nature/Gabriel Dover.
        p. cm.
    Imagined correspondence of the author with Charles Darwin.
    "First published in Great Britain in 1999 by Weidenfeld & Nicolson"—T.p. verso.
    Includes bibliographical references (p.).
    ISBN 0-520-22790-5 (cloth: alk. paper)
        1. Evolution (Biology) 2. Molecular evolution. 3. Natural selection. I. Darwin,
    Charles, 1809-1882. II. Title.

QH366.2.D67 2000
576.8—dc21   00-042615

Printed in Great Britain by
Butler & Tanner Ltd, Frome and London

9   8   7   6   5   4   3   2   1

CONTENTS

The twelfth correspondence   148
HOX! HOX! HOX!

The twelfth correspondence continued   183
TWO MINUTES THAT SHOOK THE WORLD

The thirteenth correspondence   197
BORN TO ADOPT

HOW TO READ THIS BOOK

I've been carrying this book around in my head for ten years and left myself ten days to write it (give or take a few months). Ten days that shook the world? Such random associations involuntarily enter my mind – the Grand Delusion has an edge to it. Why do I mention this? Is it to be blunt and honest about what every author secretly aspires to? Is it to communicate objective science through my subjective self-consciousness, knowing that the two are intimately bound together: the greater the truth of the former, the larger the rewards to the latter?

Igor Stravinsky was once asked how he came to compose *Le Sacre du Printemps* – that shuddering depiction of the evolutionary forces of sex and sacrifice that changed our twentieth-century perception of the point of music. In his own inimitable way and through heavily accented English, he replied with gravitas: 'I am the vessel through which *Le Sacre* passed.' The reply was the height of disingenuousness and brilliant for it. Of course, Stravinsky was proud of his creation – at a stroke he had so stepped out of the accepted bounds of society's musical conventions that he was prepared to disown it and let it merge with the elemental forces of nature.

My book is no *Le Sacre* and I'm no Stravinsky. But I am responsible for an attempt to bring to life, in a personal way, some of the more dramatic findings of late-twentieth-century biology. These are depicted through a lengthy correspondence between myself and Charles Darwin, and should, I believe, change our perceptions of the forces that have shaped biological evolution, including our own human condition. This is a tall claim. It requires a degree of immodest conviction and the ditching, stylistically, of the false 'seriousness' with which scientific writing is normally choked. Scientists have to be clear about what they mean. Darwin did precisely that in his carefully chosen narrative style of *On the Origin of Species*. There is something of Stravinsky's 'vessel' in Darwin, of which Darwin was not unaware.

The idea of a correspondence between myself and Darwin came to me over a year ago in 1998. I recognized the usefulness of this ploy to explain the evolutionary significance of modern biology and the 'new genetics', from the first elementary steps through to quite advanced levels of understanding, given that Darwin, as the father of evolution, was the most natural target for my letters. It is not difficult to imagine that Darwin would have welcomed and understood all this new information with the same zeal with which he became famously curious about all things biological in his own day.

The pace of major discoveries in biology and genetics over the past forty years has had a twofold effect, one positive and one negative. On the bright side, our current experimental and analytical techniques for uncovering some of the deepest secrets of biological organization and evolution are nothing short of miraculous compared with the endless description and cataloguing that went in the name of biology, even within living memory. Biology has moved from the static to the dynamic. The downside of this exhilaration is that few biologists have the time (or inclination?) to think through what it might all mean. The study of biology has become shattered into mutually incomprehensible subdisciplines, with their own jargon, concepts and heroes. It takes a brave biologist to attempt to draw together the nuts and bolts of molecular biology, genome dynamics, developmental processes and population biology – and all under the protection of the umbrella of evolution theory. Most biologists are more likely to implode on their microspecializations than they are to provide a wider synthesis of seemingly disparate data. Indeed, almost all of my examples in the letters are drawn from the world of animals, despite the fact that I began life as a plant geneticist. My only defence is that the world of plants is so beautifully and imaginatively dealt with by my long-time colleague, Enrico Coen, in his book *The Art of Genes* that any attempt by me to cover the same ground would appear crude and superficial.

My professional interest in biology and genetics has always been rooted in evolution. I have no explanation for this yen. It has been said that nothing in biology makes sense except in the light of evolution, to which we can add today that nothing much in evolution makes sense except in the light of genes. For it is ultimately only through the twists and turns of the genes, as they relate one to another and all to the environment during the construction of an organism, that some understanding might emerge as to why biological organisms are the hopeless, yet successful, messes that they are.

It has not been my intention to write an encyclopaedic coverage of all recent excitements in modern biology. Such coverage would have been an exhaustive undertaking, lacking any real message of interest to Darwin. I am interested more in the processes than in the patterns of evolution. Darwin himself was well aware of the crucial difference between pattern and process. His overriding concern was to propose the natural selection 'process' of evolution by marshalling the most useful and relevant 'patterns' of information in support of his claim. In a similar vein, I have preferred to select particular 'stories' from the treasure-trove of modern biology that illustrate why I believe there are several processes, not just one, that govern evolution. These processes make evolution far more complex and intriguing than can be expected from Darwinian natural selection acting alone. As Darwin said of his own *On the Origin of Species*, it represented 'one long argument'. I have attempted to do the same. I engage in one continuous discourse with Darwin in the hope of developing a more comprehensive and up-to-date perspective on the same subject: the origins of species and biological novelties.

The recent discoveries in biology are challenging how we think about biological evolution. We are entering a world of biology that lies far beyond the naivety of selfish genes and their supposedly lonely pursuits of self-replication and self-immortalization. I hope to show that the central feature of evolution is one of tolerance and cooperation between interacting genes and between organisms and their environment. Nature is not solely an arena for competitive selfish replicators, even though everyone is busy looking for the next meal and for the next mate. In all probability there has been, from the origin of life onwards, an intimate interaction between consenting genes and proteins as they learned the trick of producing successful, autonomously reproducing cells and organisms. Genes are born to cooperate.

To have a theory of evolution we need a theory of development; but to have a theory of development we need a theory of molecular interactions during the construction of an organism. We don't have a theory of interactions and so it is difficult to have a comprehensive theory of evolution. Indeed, I shall argue that we can no more have a theory of developmental interactions than we can have a theory of history, despite the attempts to produce one. So can we really have a true theory of evolution? Both individual development and evolution are the result of chance, ungoverned by any 'laws' of nature.

Down among the molecules, flexibility and tolerance permit the continual production of new life forms while retaining life's essential functions. The evolutionary process of life is akin to the trick of continually improving the structure and function of an aeroplane while the machine is in the air. Life does not have the luxury of being down on Earth's drawing board. New discoveries of the structure and behaviour of the genetic material reveal how evolution might be achieved. High redundancy and turnover of genetic elements coupled to the modular construction of organisms allow essential biological functions to be maintained during evolutionary change. Natural selection has a central role in ensuring that interacting molecules are continually in step during the incessant diversification of biological activities.

But there is more to the origins and spread of successful biological novelties than natural selection. I take Darwin by the hand and lead him through the evidence that the continual turnover of genetic elements, particularly those involved in controlling when and where a gene is used during development, leads inevitably to the 'molecular drive' process of evolutionary change. Like natural selection, it is a way of spreading a genetic novelty through a population, with the passing of the generations. But its modus operandi is very different from natural selection. As a result, new forms can arise through a combination of internal pressures generated by the unruly genes (molecular drive) and external pressures generated by an unruly environment (natural selection). Darwin focused on the latter mechanism, as the former was not known to him. But I believe that Darwin would have been the first to appreciate that the study of evolution has moved on from Darwin, just as the study of physics has moved on from Newton.

So, let me say something about the writing of this correspondence and how I would like you to approach it.

Professional scientific writing is invariably shy and constipated.

Popular commentary on science is frequently out of date, melo-dramatic and wrong. Some of the worst popularists are currently having a field day and professionals are happy to get a mention in despatches. Between the two, little useful seems to be achieved by way of a more meaningful and widely understood synthesis, rooted in fact. Out in the public domain, the culture of science remains depressingly superficial. Genetics in particular seems to get the brunt of the mischief.

I have no magic wand to wave to improve this situation. But the writing of letters is a well-understood and popular format (recently resurrected with the advent of electronic mail). I thought it might prove a useful way to take you the readers from simple beginnings to complex understanding without resorting to jargon, while permitting the two correspondents to engage in a degree of spontaneity and personal asides.

That said, I beg your indulgence over certain unavoidable restrictions. First, I cannot hide the fact that I wrote both sides of the correspondence. Nevertheless, I have used Darwin's replies as a way of asking questions on his and your behalves as the narrative unfolds. Such replies start tentatively but then increase in substance as time goes on. Second, I have made no attempt to compose Darwin's letters in mid-nineteenth-century English. I am not capable of doing that, nor do I think it is relevant to my main concern to tell and discuss the story of modern evolutionary thinking. Third, it would have been exceedingly tiresome to stop the narrative at each and every entry of modern-day terms such as computers, aeroplanes, electricity, rock music and football in order to sustain the belief that I really am conversing with a nineteenth-century scientist. I am assuming throughout that Darwin has kept up with modern gadgetry, if not with modern biology.

I hope that the advantages of communicating in the form of letters outweigh any disadvantages. I have been able to write with some freewheeling spontaneity of expression that you might find appealing and easy on the eye. We all like to engage in a little bit of eavesdropping on other people's correspondence. Indeed, as with real letters written in real time, my correspondence with Darwin has the freedom to meander in tangential directions as the subject matter dictates. I have made no great attempt to be perfectly logical in the

order of appearance of different items of interest. Instead I have tried
to give Darwin an opportunity to ask all the questions he desires,
even if some of these relate to matters yet to be introduced in later
letters from me.

My writing of the letters follows closely on what, in a sense,
Darwin said of his own writing methods: 'Formerly, I used to think
about my sentences before writing them down; but for several years
I have found that it saves time to scribble in a vile hand, whole pages
as quickly as I possibly can, contracting half the words; and then
correct deliberately. Sentences thus scribbled down are often better
ones than I could have written deliberately.' I have to confess that
without the ever-smiling presence of my secretary, Samantha
Buchanan, who can read my mind as much as my 'scribbles in a vile
hand', I could not have written and rewritten what is now before
you.

Despite the letter format and spontaneity, there are some difficult
passages to get through. As with any other discipline, these are
unavoidable. In fact in some places, I have to confess, I have written
down the hard, cruel, complex facts just to prove to Darwin what a
'mess' biological organisms have evolved into. I have used diagrams
where the going gets tough, and have included a glossary of unfamiliar terms. But I warn Darwin where these sections are, so that the
principles of what I am saying, rather than the precise details of this
or that biological phenomenon, form the basis of our discussion.
This is back to the issue of 'process' and 'pattern'.

Darwin is, of course, a real person and not a figment of my imagination. Accordingly, I have not shied away from making personal
asides about his life and bringing in personal aspects of my life that
could be regarded as irrelevant and self-indulgent. But, as the song
goes, 'it's my party, and I'll cry if I want to'. It has been enormous fun
writing these letters and I hope you get into the spirit of them. There
are flashes of humour here and there, but they are intended simply as
breaks in the very serious business of relaying to you the nature of
modern biology and how we think it might have evolved at the turn
of the millennium. I believe that this discourse should affect how we
view ourselves as evolved biological beings, participating in the
illogical yet coherent living world around us. I try to offer a running
commentary, especially towards the end of the correspondence, of

the significance of the new genetics to issues of human individuality, genetic determinism, race and free will.

During the past ten years I've done a lot of talking – some might say too much. But chewing the fat and shooting the breeze with a host of colleagues and good friends have been essential for understanding the new biology. I have been influenced in a variety of ways by some key raconteurs and correspondents, even if some of them are unaware of the fact. I particularly want to mention Antonio Garcia-Bellido, Enrico Coen, Eric Davidson, Steve Gould, Dick Lewontin, Gerry Edelman, Alec Jeffreys, Alfonso Martinez-Arias, Sydney Brenner, Francis Crick, Dick Flavell, Pat Bateson, George Miklos, Bambos Kyriacou, Ed Southern and the late Allan Wilson.

Samples of Darwin's signatures and handwriting were graciously provided by the Wellcome Library for the History and Understanding of Medicine. Financial support from The Leverhulm Trust and Weidenfeld and Nicolson has been usefully spent: long may it keep flowing. I'm very grateful for the professional advice and support of my editor, Peter Tallack, throughout. He saved you from a further twenty-five thousand words!

This book of letters is for my daughter Merav and my two sons Noam and Alexis, and for my daughter's son and daughter, Micha and Manya, all of whom I hope have learned, and will continue to learn, more from my nurture than from my nature. Or is it the other way round?

Gabby Dover
Oxford

THE TWIN PEAKS

Date: The turn of the millennium

Dear Mr Darwin

You might find it presumptuous of me, if not a little macabre, that I should take up my pen and write to you more than a hundred years after your death. But I'm encouraged to do this because it is on record that you yourself wrote almost fourteen thousand letters on scientific issues, many of which I expect were answers to unsolicited correspondence. By the standards of today that is a tidy pile to get through. Indeed, contemporary letter-boxes are more likely to be filled with commercial advertisements than hand-written letters leisurely composed between correspondents pursuing a shared interest.

Be that as it may, I've become bold enough to write to you because you are recognized as the father of the study of evolution. I cannot know whether or not, deep in your tomb in Westminster Abbey, you've been keeping abreast since your demise with the ups and downs of your theory of evolution, which you called, quite cleverly (if a little misleadingly as it turned out), 'natural selection'. Indeed, I'm not sure whether my modern English usage, directness of speech and lack of etiquette as a late-twentieth-century scholar will be understood or well received by you. But it is not just the words that have changed. The current world of learning is very different from the one you inhabited in the nineteenth century, particularly from that very private perspective of your country retreat at Down House in Kent, over a forty-year period. I envy the solitude you must have enjoyed as you probed into the best scientific minds of your generation through your correspondence, and I covet your ability to rope your tireless wife and large family into helping with your intriguing home experiments. Your world of an inspired self-funded scientific

gentleman is a far cry from today's great unwashed masses of professional scientists competing 'red in tooth and claw' for a small pot of
experimental funds and scientific credibility.

Despite the gulf that separates us in time and means, I'm hoping
that this letter will arouse your scientific interests, for it touches on
some of the central issues in the theory of evolution with which you
wrestled all your life. I will even be so bold as to say that your interest *will* be aroused, not just because of the very human curiosity I
expect you to have about what we think of you a hundred years later,
but also because I have a story to tell of modern genetic wonders that
should open your eyes.

The story I will relate adds substantial flesh to the bones of your
theory of natural selection, which should please you immensely.
However, by illustrating the bizarre antics of genes both in evolution
and in the development of organisms, I hope to reveal that there is in
fact more to life than natural selection. I'm confident that the same
intensity and honesty that drove your scientific curiosity to seek out
answers to the big questions of the origins and diversity of life will
lead you thirsty to the well of our late-twentieth-century understanding of the forces that have shaped life's evolution. I hope that
you will take the opportunity to write back to me with your
questions and comments as you see fit.

The Leicester connection

Mr Darwin (I don't dare call you Charles, even though first-name
terms are *de rigueur* in today's libertarian society), first let me introduce myself. I am an investigator of genetics and evolution at the
University of Leicester, a city that has the distinction of providing
Alfred Russel Wallace, the co-founder of the theory of natural selection, one of his first jobs as teacher of classics in the local college. It
was in Leicester's public library that he accidentally met your friend
and correspondent, the naturalist Henry Bates, and in 1848 the two
young men set off on their own epic voyage to the Amazon, armed
with copies of your *Journal of the Voyage of the Beagle* and Thomas
Malthus's *Essay on the Principle of Population*. I'm a little embarrassed to remind you how your life's work almost crumbled around
you on that fateful day, ten years later, when you received the

twenty-odd pages of a letter from Wallace written on his sick-bed in the Spice Islands. It described his independently conceived idea of a mechanism of biological evolution by means of natural selection. In that inhospitable place, far from the comforts of London, he was struck with inspiration during one feverish fit about the evolutionary significance of Malthus's theory of overpopulation. As with you, the penny had dropped. And the rest is history, as they say. For twenty years you had been assembling the evidence and composing chapter after chapter, painstakingly and not without a little angst, of your big 'Book' of sketches on species and their transformation. Now something had to be done, and quickly if you were not to be 'forestalled', as you yourself put it.

As you well know, this was achieved in the most courteous of ways by your friends, Charles Lyell and Joseph Hooker, two leading figures in nineteenth-century geology and biology, who managed to arrange for you and Wallace to 'read' your joint scientific papers on evolution at the same meeting of the Linnean Society in 1858. Of course, neither of you was actually there – Wallace was truly suffering on his south-east Asian island and you were inflicted with an undiagnosed illness that plagued you all your life since the return of *The Beagle*. I am full of admiration for your genuinely expressed concern not to be seen as acting dishonourably in agreeing to the joint presentation, having had privileged access to Wallace's theory. Similarly, the admiring Wallace later told Hooker that he would have suffered badly had he published his paper alone, ahead of you. Although your joint announcement of the first workable mechanism of evolution left the fellows of the Linnean decidedly unimpressed, your behaviour to Wallace and his to you were exemplary. I cannot resist telling you that in today's world the course of events might have been very different. The unknown Wallace would have faxed his manuscript to the editor of *Nature*, who in turn would probably have sent the piece to Thomas Huxley, one of your fiercest supporters, or to Hooker, for peer review. One of them might very well have advised *Nature* to reject it, on the grounds that the paper was 'not of general interest' or 'much too speculative'!

But let me get back briefly to Leicester. It is also the city where one of our leading contemporary geneticists, Professor Sir Alec Jeffreys, has made major discoveries of the nature and behaviour of the hereditary material of what we call today 'genes', which are of fundamental importance to the story that I shall relate. Before my own arrival here, I spent the bulk of my scientific career as an evolutionary biologist at the University of Cambridge, whose general appearance cannot have changed too much from the one you inhabited. Indeed, I cycled almost every day past the undergraduate lodgings in Fitzwilliam Street, which bear the plaque 'Charles Darwin lived here'. Heavy-metal rock music might have replaced the clatter of horses' hooves, but the spirit of the place hasn't changed. You have cast a long and abiding shadow on the world of learning in all its ramifications, perhaps beyond your wildest dreams. Even the house that belonged to your descendants is now Darwin College, specializing in postgraduate studies.

Your theory of natural selection is being increasingly exploited as the explanatory cornerstone of almost any unsolved mystery in human biology and affairs. For example, major new disciplines have sprung up in evolutionary psychology, Darwinian medicine, Darwinian morality and ethics, and Darwinian linguistics, in addition to the traditional spheres of biology, ecology and the environment. As the perceptive Russian-American evolutionary biologist Theodosius Dobzhansky has written: 'Nothing in biology makes sense except in the light of evolution.' This aphorism has recently been taken to heart with a vengeance and with a good deal of weak evidence and excessive speculation.

Taking your name in vain

Indeed, I'm not at all sure whether you would be entirely happy with what passes as scientific evidence for the supposed role of natural selection in our human attributes or in the strange things that animals, plants and microbes get up to. For example, there is a widespread assumption that the very existence of any given, successfully functional, structure, such as the eye, is proof enough of the workings of your theory of evolution: 'I function, therefore, I was naturally selected'. The assumption here is that any set of biological parts that work smoothly together must have arisen by natural selection, for there is no other way in which such complex functions could have arisen. Armed with this assumption, the provision of scientific evidence for the role of natural selection has somehow got lost on the way.

I intend to convince you that the evolution of any complex organ, like the eye, or human language and consciousness is an intricate and fascinating mix of a variety of evolutionary activities, of which natural selection is one. There is, in our new understanding of biology, no dismissal or diminution of natural selection, but a healthier regard for the need to provide scientific evidence for its role *per se* and for the important interaction it has with the newly discovered activities of genes. As I shall show, genes have a mysterious life of their own that can greatly influence the course of evolution.

Sublime yet pathetic

If I were to try and summarize for you what statement encapsulates the essence of your great insight into the nature of biological organisms, it would be your recognition that there is something both sublime and pathetic about them. In thinking deeply about the lives of organisms, about how they construct themselves from a fertilized egg and how they reproduce themselves, you recognized the dual play between chance and necessity in putting it all together. There might, indeed, be a workable and successful function called the human eye, but the actual organ in front of us is by no means the only one that might have evolved for the act of seeing. It happened to be the one that emerged from the single, particular path of evolution that actually took place, given the available starting materials and the series of unique chance events along the way. But there could have been many different routes to similar effects had evolution taken different courses. The particular evolutionary paths that were taken have not necessarily produced the most efficient, streamlined machine for seeing: the eye, for all its sublimeness in coordinated sight, is rather pathetically wasteful of energy, is too complex and has too many moving parts. It's a veritable dog's breakfast of evolution and development.

Our problem is that an organ such as the eye is so unnervingly familiar that we regard it as 'natural', in opposition to all the 'unnatural' organs that were supposedly not allowed to happen. But I believe that there was no absolute logical necessity to the particular

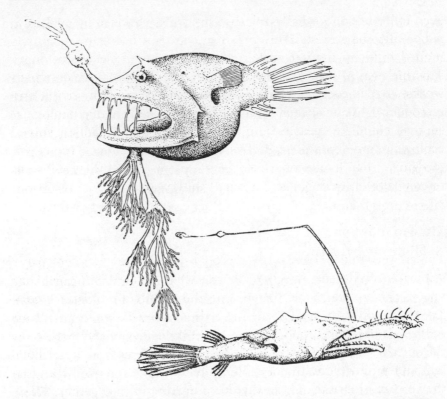

Natural or unnatural? *Unless you are an expert in deep-sea angler fishes, it is hard to decide if these are real or imaginary likenesses. We tend to regard the unfamiliar as unnatural. You might find it even harder if I asked you to provide an explanation for why the structures have evolved as naturally selected adaptations. Finally, you might throw your hands up in despair if I tell you that these two fishes appeared in the book* Animal Biology *written by J.B.S. Haldane and Julian Huxley (the grandson of your friend Thomas Huxley), two of England's leading students of evolution in the middle decades of our century. Several colleagues of Haldane and Huxley were quick to point out that the fishes had an uncanny resemblance to the book's authors, neither of whom you would want to run across unexpectedly in the darkness of some deep argument. Is there a natural-selection explanation for such resemblances? In this case, no. But unfortunately, there seems to be no limit to what 'scientific' imagination can produce by way of adaptational tall stories in the bizarre world of biology.*

set of biological processes governing the organisms that happened to evolve on our singular Earth.

It is my belief that you recognized this deep fault-line running through biology and understood this tension between the sublimeness of function and the pathetic nature of evolved structures, which is at the very heart of evolution by natural selection. Today we have even more reason for believing that anything produced by evolution is bound to be a mess. Life's organisms are not lean gene-machines fuelled only by natural selection, purring down the evolutionary superhighway. What is important in evolution is that an organism works; that it goes about its daily business of growing, living and reproducing. Your special insight was to recognize that biological success could be measured in terms of how an organism solved ecological problems affecting its ability to reproduce. Today we recognize that evolutionary success can also be measured with reference to other important criteria that I'm eager to relate to you.

Time to move on

You are recognized, like Isaac Newton, as one of the most perceptive scientists of all time, in this or in any other country. But I feel that you would be one of the first to appreciate that evolutionary theory has moved on from you, in the same way as physics moved on from Newton. Newton's principles about the forces governing motion on the surface of the Earth are not subtle enough to deal with our contemporary understanding of the origins of the atoms and energy in the Universe, or of the forces that keep together the subatomic particles inside atoms. This is not to say that Darwinian selection or Newtonian mechanics is misconceived; it is a statement that recognizes the existence of many other contributing factors to biological or physical phenomena that neither you nor Newton could have known about, or ever guessed at. Both of you, I'm sure, would have been comfortable with the new biology or the new physics, had this information been available to you. Both of you might have produced suitably comprehensive theories for our admiration. Your intellect did not stop at natural selection because that's all it was capable of producing. You were a child of your time, as we are all children of our time. As an honest and creative scientist, you would not allow yourself to venture too far beyond the realities of biology as they were known in your day; I am sure that, if you were alive today, you would be like the little boy lost in the proverbial sweet-shop. So how can I begin to whet your appetite?

'Natural Sorting'

I intend to use the rest of this introductory letter to ensure that we are talking the same language about your theory of natural selection and to explain the laws of inheritance that were unknown to you. These laws were eventually deduced by a scientific monk working in his monastery garden in what was the Austro-Hungarian empire, but is now the Czech Republic. At the risk of being accused of teaching my grandmother to suck eggs, I want to start my story in what is all too familiar territory for you.

Like most of your contemporaries, you had no concept of the nature of a gene; even the word itself had not yet been coined. What you did know, however, as one of countless parents who had ever wondered why their children were more likely to bear a resemblance to them than to unrelated people, is that some material 'thing' was passed on during the sexual act from one generation to the next. This 'thing' is what we now call collectively the genes. It was your great insight to propose that the inherited genetic differences between us all could serve as grist for the mill of natural selection. Your argument was simple in the extreme: should any of the inherited differences influence how many offspring some individuals might have relative to other individuals, then clearly a greater proportion of the next generation would have inherited the genes that had contributed to the differences in numbers of offspring.

You recognized that an individual's ability to survive and reproduce relative to its peers is an outcome of how well that individual's features are adapted to a given ecology: that is, its relationship with other organisms and to the physical environment in which they live. The genes simply ensure that some of the reasons for better survival and a more successful rate of reproduction are passed on to the next generation. If, then, at the second generation, the same set of individual features for better survival and reproduction are at a premium, the same genes responsible for such features would populate the third generation even more. Eventually, after hundreds of thousands of generations, a particular genetically influenced adaptation would evolve. In other words, there would be a gradual accumulation of the biological features that improved survival and reproduction,

alongside a gradual accumulation of the genes responsible for contributing to these features during the growth and development of
each new individual at each new generation.

You called this process 'natural selection', bearing in mind the
commonly understood practice of artificial selection as practised by
breeders of dogs, horses, pigeons and orchids.

Calling the process 'natural selection' was a stroke of genius, but
unfortunately it has led to 150 years of confusion. I will devote a separate letter to this issue, but let my say here at the outset: as you
know, natural selection is not an 'active process' like artificial selection. The disproportionate increase of some genes at the expense of
others, as one goes from one generation to the next, is a passive
outcome of the particular interactions between prevailing ecological
conditions and a particular set of genetically unique individuals.
From our modern-day understanding of what happens to genes
during their passage from one generation to the next, we can state
categorically that the new individuals that arise at the next generation will also be genetically unique. In sexually reproducing species,
no individual could be the same as any individual that ever lived
before, or be the same as any individual that is yet to come. The
probability of two sexual individuals being the same in the course of
evolution is so infinitesimally small that it can be safely ignored,
excepting genetically identical twins.

For the moment I am confining myself to the genetic contribution
to individual uniqueness. The environment also uniquely influences
the development of individual form and behaviour, but I am ignoring
this for now. As a result, to return to natural selection, the nature of
the interaction between a new set of genetically distinct individuals
and the local ecology will be different from the interactions that took
place in the previous generation. Furthermore, the subtle features that
define a given ecology will not necessarily be the same in the next generation. The relationships between organisms and their ecology that
govern which individuals survive and reproduce more successfully
will be specific to each generation. It is not legitimate, therefore, to
make predictions about the continuing, selective success of specific
genes, generation after generation, as if selection, for or against a gene,
were a constant property of the gene. There is no active agent of
selection: there is no Great Selector in the sky. 'Selection' is a one-off

outcome of a unique set of relationships between a unique set of individuals and a unique set of ecological conditions.

Much of the confusion might have been avoided if you had named your mechanism 'natural sorting' or 'natural filtering'. This would have implied a differential sorting of genes at each generation owing to the reproductive successes of unique individuals. There is a difference between selection and sorting. Imagine you had a mixture of wheat grains and wanted to separate the small grains from the large grains. You could either select all the small grains by hand or allow the small grains to sort themselves by shaking the grains over a suitable sieve. If the sieve represents the particular set of interactions among individuals and their surrounding ecological conditions in any given generation, then the size and number of sieve holes will change in all succeeding generations.

I am sorry to belabour you with this obvious point, but it is one of the sad features of current trends that genes are assumed to have uniquely identifiable and constant selective 'powers'. Their ability to propel themselves through evolution is regarded as a mark of their selective selfishness. I'll discuss more of this in later letters when I'll describe to you some recent surprises about how genes affect the development of organisms and a range of unexpected gene antics. When I've finished this story about our new perceptions, you will see the extent to which the individual organism, rather than the gene, returns firmly centre stage in the mechanism of natural selection and in any other evolutionary process. But before I can properly advance any further on this journey of discovery, I need to tell you the story of how the genetic mechanism of inheritance was revealed just a few years after you wrote *On the Origin of Species*.

The monk in his garden

Your theory of natural selection did not need a precise knowledge of genetics to be formulated properly. But it did suffer from not being able to state how individuals came to differ and how such differences could be inherited. Although you had some very powerful and influential supporters of your theory, most biologists were uncomfortable with natural selection right up to your death in 1882 because the mechanism of inheritance had not been resolved.

It is one of the great ironies of history that the problem was resolved in your own lifetime by Gregor Mendel, who published a paper of some forty pages in the *Proceedings of the Natural History of Brünn* (now Brno, Czech Republic) in 1865 that was to revolutionize the study of biology. It is believed that you might have seen a mention of Mendel's studies of inheritance in plants as early as 1874, in a book by Herman Hoffman. Six years later you received a book from another German botanist, containing a much longer citation of Mendel. This book was not read by you, as the pages were found to be uncut. What has remained a mystery is the reason why you provided Mendel's name for inclusion in an article on plant hybridization for the Encyclopaedia Britannica in 1881. Had you developed a sixth sense as to the importance of Mendel in our twentieth-century understanding of biology and evolution? This sounds like a nice story, but I don't think that is the case, alas!

Unfortunately, Mendel's paper came to general notice only in the early years of the twentieth century. I think that you and Mendel would have enjoyed each other's company. You were both reclusive – you in Down House and he on his chosen path as a monk and abbot in the monastery of Brno – and you were both involved in interfering with the natural reproduction of plants, in the best possible taste.

The great foresight of Mendel was to work out the rules governing the inheritance of genes, even though, like you, he had no knowledge of the substance of a gene. In his monastery garden he would take the male pollen of one type of pea and use it to fertilize the female stigma, and ultimately the egg, of another type of pea. He chose pea plants that differed in the colour and shape of their seeds. By counting the numbers of offspring that have one or other of the colours, and one or other of the shapes, he realized that the underlying genes were being distributed among the next generation according to the laws of chance. For example, if you tossed two coins together 100 times, on roughly half the occasions the two would fall as one head (H) and one tail (T), a quarter of tosses would show two heads and one quarter would show two tails. This same 1:2:1 product of chance (HH:2HT:TT) applies independently to each of his pea genes, one controlling colour and the other shape.

Mendel's flash of inspiration was to deduce from this that the cells making up each individual pea plant contained two copies of each gene, like the two sides of a coin, but that only one copy entered each pollen grain or each egg. When a pollen fertilized an egg, the new individual is reinstated with two of each gene. Naturally, only genetic alterations in the sex cells of pollen or eggs are of evolutionary importance because they are the only ones passed on to the next generation.

One of the more amusing veranda conversations I have with friends is to discuss whether you would have realized the evolutionary significance of Mendel's results had you yourself carried out the pea-crossing experiments, or whether Mendel would have been similarly perceptive about the evolutionary implications of his results had he learned about natural selection. My suspicion is that both of you, in your separate ways, had achieved enough for one lifetime: history tells us that it took the combined forces of several leading theoretical biologists and mathematicians during the first thirty years of the twentieth century to recognize the enormous fillip that Mendel's rules gave to your theory of natural selection. But before I move on to this, I can't resist reminding you that your first cousin, Francis Galton, wrote a letter to you in 1875 in which he described the 1:2:1 probabilities of genetic segregation – but, alas, he failed to see the whole picture. Just think, you had the chance to lay the foundation stones of both evolution and genetics, and keep it all in the same family.

So, what is it about the peas in the Brno monastery that formed the cornerstone of what became known as the 'neo-Darwinian synthesis': the arranged marriage between Mendelian and Darwinian concepts that was consummated in the 1930s?

The genetic lottery

To answer this I need first to explain why Mendel's rules of genetic inheritance obeyed the same laws of chance that govern the tossing of pennies and dice. It is all to do with chromosomes. These are wriggly worm-like objects in the nuclei of animal and plant cells, viewed under the microscope by many of your contemporaries. Each individual chromosome can be viewed as a long piece of string (composed of a macromolecule called DNA – but more of that later) in which certain sections of the string are the genes themselves. An individual chromosome can contain several thousand genes.

As you have no doubt guessed from what I just told you about the existence of two copies of each gene in an individual's cells, there are also two copies of each type of chromosome. For example, in each human cell there are twenty-three pairs of different types of chromosome, and each member of a pair is inherited from either one of our two parents.

A single human sperm or egg contains only one set of twenty-three chromosomes, and after fertilization the cells of each new individual contain two sets of twenty-three chromosomes. The sex cells are said to be haploid and the body cells are said to be diploid. Given this alternation between one set and two sets, there has to be a mechanism that accurately separates the two copies of each type of chromosome during the formation of sperm or eggs. This is the job of a process of cell division called meiosis. The nub of this process is to line up the twenty-three pairs of chromosomes along the middle of the nucleus and then to pull one set to one pole of the cell and the other set to the opposite pole. It is at this juncture that one of the most important events in biological evolution takes place: each pair of chromosomes can line up and present any chromosome it wishes to face a given pole of the cell, independently of the orientation of all the other pairs. In other words, the set of twenty-three chromosomes that we inherit from our father does not trundle off as an intact set to one of the poles; nor does the maternally derived set go to the opposite pole. Instead, each pole ends up with a random mixture of the original paternal and maternal chromosomes. The only constraining rule is that one chromosome of each type of chromosome has to be present at each pole, so that there are twenty-three different types of chromosome in each sperm or each egg.

This shuffling of chromosomes ensures that the two sets of twenty-three in the first cell of the new individual are composed of a random mixture of the chromosomes of the four grandparents and the eight great grandparents and so on, all the way back through the generations.

The endless shuffling of chromosomes during the formation of eggs and sperm, coupled to the random process of fusion of any egg with any sperm, is of great importance for your theory of natural selection. Mendel's rules of genetic inheritance are an outcome of the behaviour of chromosomes at meiosis and during fertilization.

Although Mendel was observing the behaviour of genes that controlled the colour and shape of pea seeds, the genes segregated
according to the same laws of chance that stem from the behaviour
of chromosomes during the sexual process. This is because, at the
level at which Mendel was making his observations, the behaviours
of genes and chromosomes are considered to be one and the same
thing. Indeed, from what I said earlier about chromosomes containing the long piece of string (DNA) that constitutes the genes, you
may interject that it is to be expected that wherever the chromosome
goes, the gene follows suit.

At the risk of confusing you, just when I'm getting you up to
scratch about the simple rules of inheritance of genes and
chromosomes, I will say now that one of the major surprises of the
study of genes in the latter half of the twentieth century is that the
behaviour of genes and chromosomes is not always the same thing.
They are slightly out of phase with each other. This is a small,
almost imperceptible, difference in the patterns of segregation that
would not have been noticed by Mendel, or indeed by hundreds of
subsequent geneticists involved with crossing plants and animals.
On the time scale of evolution that concerns you and me, however,
the essentially 'non-Mendelian' behaviour of genes would begin to
bite.

But I'm running ahead of myself. First I need to tell you that,
assuming that genes and chromosomes are in perfect synchrony
during the sexual process (a reasonable assumption in the 1930s),
Mendel's rules of inheritance finally convinced biologists that your
theory of natural selection was of absolute necessity if we are to have
a realistic and rational explanation for the improbable origins of
biological novelties.

Let me explain why this was so.

The case against the hopeful monster

One of the most important phrases that you penned about evolution
was that it is a process of 'modification by descent'. In other words,
all evolutionary processes, whatever their precise mechanism of
operation, need to bring about a gradual modification in some feature
of an organism with the passing of the generations. In today's

terminology, we see evolution as a gradual change in the average genetic composition of a population of individuals. Notice here that I've switched from considering a modification of one individual to that of a population of several individuals which we will assume are sexually reproducing. We cannot talk about one individual because we then enter the 'hopeful monster' dilemma that caused some amusement during the middle part of the twentieth century. For example, let's say that, by some miracle of genetic mutation and chromosome shuffling, one monstrous long-necked individual (our first giraffe) arose in a population of the ancestral short-necked giraffes. It is highly unlikely that the genes that contribute to giraffe neck development could work successfully in any offspring of this mutant long-necked giraffe and a short-necked giraffe. The two sets of genes that need to coexist in the nucleus of the new fertilized egg would probably have two contrasting influences on neck length, and this is bound to cause developmental confusion.

There was a short period in the history of evolutionary genetics when some serious thought was given to the role of gross mutations and the one-off production of 'hopeful monsters' during evolution. Such scenarios are no longer considered tenable because they violate your phrase 'modification by descent'. The secret to an evolutionary process is to show how there is a gradual change in the average genetic constitution of individuals so that there are no overt disturbances in the development and reproduction of individuals in a population. Natural selection, naturally, sits comfortably within your central requirement. So do the processes of neutral genetic drift and molecular drive, which I shall come to in due course.

Mendel to the rescue

We can, therefore, eliminate gross one-off mutations from playing a major part in the origin of species and biological novelties. But what about Mendel's rules of genetic inheritance? Is it possible that the patterns of segregation could bring about a gradual 'modification by descent' as genes are passed on from one generation to the next?

In 1908, within several years of the rediscovery of Mendel's results, it became apparent to two mathematicians, George Hardy and Wilhelm Weinberg, that the answer is a resounding 'no'. On the contrary, rather than lead to a gradual change, the laws of inheritance lead to genetic stability – the very opposite of what evolution needs to achieve in its incessant drive to diversification.

To explain this conclusion, I need to get you to think back to what I told you about the two pennies tossed in the air. Imagine each penny represents a particular gene in one parent, with the head and tail faces of the penny symbolizing the two copies of that gene in that parent. About half the sperm of a father will contain a head and the rest will contain a tail. Similarly, for a mother there will roughly be a 50:50 distribution of head-containing and tail-containing eggs. Now, given that any sperm can fertilize any egg, as I explained before, half the offspring will contain a head plus a tail (HT), a quarter will contain two heads (HH) and a quarter two tails (TT). We immediately have three types of genetically distinct individual, HH, HT and TT, in the proportions 1:2:1.

Now let's consider a further round of mating. If an HT father mates with an HT mother, we get our now familiar 1:2:1 ratio. But matings can also occur between an HH parent and an HT parent, or a TT parent and an HT parent, or indeed TT with TT, and HH with HH. In fact, all possible combinations of mating can occur. If, say, HH mates with HT, we get a 1:1 ratio of HH:HT offspring. HH with HH gives all HH offspring and so on. If we now add up the proportions of HH, HT and TT offspring from all the matings that took place across the whole population, having started with a distribution of, say, 1HH:2HT:1TT at the parental generation, we find that the 1:2:1 proportions have remained unaltered. There is no change in the proportions of the two different copies (represented by H and T in my example) of the gene in question – there is no 'modification by descent'.

It does not matter which initial proportions we start with at the first generation: even if, say, 80 per cent of the population are HH and the remaining 20 per cent are either HT or TT, then these proportions will not change, generation after generation, as long as there is random mating between individuals and no one is left sulking in the wings. Such long-term stability became known as the Hardy–Weinberg equilibrium.

With this conclusion that the laws of inheritance cannot bring about evolutionary change, the architects of the neo-Darwinian synthesis realized that here was solid, theoretical justification for natural selection.

Mr Darwin, seventy years after the publication of *On the Origin of Species*, you finally received the scientific credibility for your theory of evolution by natural selection that you always craved for. You had proved in your own lifetime that organisms as we know them today evolved from earlier extinct forms. This in itself was a considerable achievement, with psychological repercussions for the 'human condition' that go far beyond the biology of the earthworms and barnacles on which you made such fascinating observations. But I suspect what was of far more importance to you was to have your mechanism of natural selection accepted as a realistic, workable model for the origins of new species and their unique adaptations. You had to wait a long time for Mendel to come to your rescue, but it did finally all click into place. You and Mendel became the twin peaks of evolutionary biology. By the 1930s you sat astride the world of biology by virtue of the scientific concept that meant most to you.

Almost but not quite

Except ... it was all too good to be true. The story of evolution theory did not end in the 1930s, and many new and quite unexpected discoveries of the non-Mendelian misbehaviour of genes reveal that natural selection is only part of the story. Furthermore, the explosion of surprising findings about the role of genes in the development of organisms requires us to modify the earlier assumptions about the lonely role of natural selection in the origin of biological novelties.

I'm going to end this first letter by giving you just one example of the unexpected behaviour of genes – it should serve as a good illustration of the underlying non-Mendelian turbulence at the level of DNA in the chromosomes. It is also a fascinating story in itself about the difficulties scientists have in getting their evidence for the unexpected accepted by their peers, especially when the observations contradict long-established and, unfortunately, highly cherished ideas.

Barbara's jumping genes

The story concerns Barbara McClintock, an American geneticist who was born in 1902 just after the rediscovery of Mendel's segregation rules and who received the Nobel Prize (the highest international distinction that can be bestowed on a scientist) eighty-one years later in 1983. In the early 1940s, McClintock began a study of the genes affecting the colour of kernels on the cobs of corn plants – inauspicious beginnings for a vital discovery in genetics. As Mendel showed with his peas, different versions of the gene were responsible for colour, so that some kernels could be entirely plum-coloured and others could be white. McClintock then noticed that in some kernels, which were essentially white, there were tiny spots of plum colour. In other words, there were groups of cells in which a genetic switch had taken place, turning the expected colour of white to plum. Normally we call such genetic switches mutations: that is, an alteration in the DNA that constitutes the gene responsible for colour. It was the existence of just such mutations in the genes controlling pea colour and shape that allowed Mendel to discover his rules of inheritance. Now, from many countless experiments that geneticists have carried out over the years, it has been calculated that the natural rate of mutation is extremely slow – one event per million copies of a gene at any one time. What was truly surprising about McClintock's observations was that there were so many groups of cells with a switch in colour that the traditionally understood mutation process would have a hard time explaining their frequency. McClintock, with her intense persistence and ingenious intellect, solved this puzzle in an astonishing way. She proposed that the phenomenon was due to genes jumping from one chromosome to another!

Although McClintock was a respected and greatly admired geneticist by the time of her discovery of mobile genetic elements, her latest suggestion was so disturbing that it led to widespread incomprehension and criticism bordering on the dismissive and hostile. As McClintock herself recollected on her ninetieth birthday, 'no amount of published evidence would be effective' in silencing the critics.

The existence of genes that can literally move from one place to another is a phenomenon that we now know happens in all forms of microbes, animals and plants. Such unexpected mobility naturally undermines Mendel's rules that genes and chromosomes always behave in synchrony during the sexual process of meiotic cell division. If bits of DNA can jump from one chromosome to another, the behaviour of genes and the behaviour of chromosomes are not one and the same thing. The implications of this are profound – it means that the non-Mendelian mode of inheritance of genes, as opposed to the Mendelian mode of chromosomes, can alter the long-term genetic composition of a population of individuals, as I shall shortly reveal.

A sneak preview

Not only can bits of DNA move around the chromosomes, but there are half-a-dozen other ubiquitous mechanisms of essentially non-Mendelian rearrangements at play at the level of the genes. So the genetic material is in a continual state of flux. Some of these mechanisms can make hundreds of copies of the same gene arranged head to tail like carriages of a train. Most genes, including the important genetic elements that control when genes are on or off, are repetitive at one level or another, and the several copies of a gene can exchange DNA between them, as if they were talking to each other. These discoveries of the late twentieth century and their implications for the processes of evolution and development will be the subject of my future letters.

But where, I hear you asking, is the role of selection in all this? Well, I believe that selection has a rich and varied part to play in counteracting the wayward tendencies of genes. And it does so not by declaring an embargo on new genetic rearrangements, but by ensuring that compensatory genetic changes take place to maintain essential biological functions. The non-Mendelian activities of the genes, coupled to selection, have consequences for our views on the origin of species, the establishment of novelties in the form and behaviour of organisms, and the tempo and mode of evolution from the origin of life onwards. The nature and misbehaviour of the genetic material also has important implications for how an individual develops from a fertilized egg. As you will hear shortly, quite complex organisms are built like a child's Lego set from a few basic units shared by all species. But the story cannot be rushed. I feel I've said enough for the moment and outstayed my welcome.

I hope, sir, that this incursion into your forced seclusion has not been too alarming.

Ever your obliged

Westminster Abbey

My dear Dover

I am so glad you have taken the time and trouble to write to me. It is one of the saddest aspects of human existence that, as soon as one passes away, it is generally assumed that the deceased has no further interest in what he or she spent a great part of life investigating. From what you tell me of the Darwin industry of scholars in your day, busy seeking out every nuance of my life and thoughts, I have to conclude that there is indeed life after death. Being human, I am pleased with the accolades that have come my way over the years, although I'm disturbed to hear that my name or, more accurately, my natural selection theory, can be accepted in some quarters without much in the way of experimental or observational evidence. We used to have the same problem in my own day, but we could at least hide behind the excuse that we were hampered by the limitations of the investigative tools at our disposal. I'm looking forward to receiving more of your letters, which you promise will provide the detailed evidence that natural selection is alive and well. Please don't skimp on the details, for I sense from your letter that your view of biology is close to my own: biology is nothing but details. Dare I say it: God is in the details. Or is it the Devil?

Given the arbitrariness of natural selection, we cannot expect that any particular evolutionary path is following some preordained set of natural scientific laws. Make-believe organisms that might seem 'unnatural' to us, when judged against the reality of the natural world, might very well have been 'natural' if evolution had taken a different historical turn of events. We need to discuss this issue more at another opportunity. However, let me just add, at this point, that physics might have its predictive rules and regulations, derived from observations of the regularity of physical phenomena, but biology, alas, is simply an historical process not governed by obvious laws.

If I may be permitted to give you an example: let us compare a population of water molecules in a beaker with a population of biological organisms. To a good working approximation, the physicist can regard the population of water molecules, at a given time and place, as being essentially the same as all other water molecules at all other times and places, once some elementary assumptions have been made about pressure and temperature. I do not need to tell you, for indeed this is one of the points you have made in your letter, that a biological population at a given time or place is a unique structure, the genetic architecture of which would never have occurred in previous generations nor is likely to occur in future generations. Notice how quickly I'm adapting to your modern-day vocabulary! Your glossary has been most helpful in this regard.

Your explanation of genes and chromosomes and what sex does to them has quite taken my breath away. I do envy Mendel beavering away in his monastery garden and making such important deductions. If only I had been a little more meticulous with the planning of my own experimental crosses, or listened a little more carefully all those years ago to cousin Francis. I recall seeing Mendel's name in various collections of contemporary scientists engaged in plant hybridization. If my German had been a little better and my page cutter a little more at hand maybe I would have become better informed. Still, Mendel cannot complain that the accolades did not finally come his way, if a little belatedly.

I do like the idea of binding me and Mendel together. As a scientist more interested in being right than being famous, I did heave a sigh of relief on reading your account of the rise of the neo-Darwinian synthesis in the 1930s. I particularly appreciate the conceptual need for natural selection to achieve my 'modification *by descent' (or what you call a change in the genetic composition of a population with the passing of the generations), given that Mendel's rules of inheritance cannot achieve this by themselves. Even though you tell me, from your perspective of the late twentieth century, that this earlier analysis is a little naive, it seems to have been an important step forward.*

The beauty of hearing all this in one go is that I'm as much open to what you are beginning to tell me about the ubiquity of non-Mendelian genes and their long-term evolutionary effects as I was receptive to your account of Mendelian genes – or should I now, more accurately, say Mendelian chromosomes. I realize that you have only just begun the telling of this story, in describing McClintock's jumping genes, but your point has been made. What I feel is needed now is for you to tell me about the other peculiar practices of genes and to give me actual examples of how they have contributed to the origin of new species. As you will appreciate, I'm just an old-fashioned nineteenth-century naturalist and I need to have real examples in front of me to appreciate what you, or anyone else, is saying about the forces that drive evolution.

The idea that organisms are modular and, as you seem to be implying, that biological complexity is about novel combinations of existing modular processes is intriguing. It makes me believe that evolution, up to and including our own brain function, is easily understandable and open to experimental dissection with your modern genetic trickery. I'm looking forward to reading evidence about this. That there is a crucial role for natural selection in all of this is pleasing to hear. It would seem that natural selection has to attend to the problems arising from the internal turbulence of genes as it does to those arising from the external turbulence of the ecology. My imagery of the 'tangled bank', which I'm sure you've read on the last page of On the Origin of Species, *will surely have to be extended inwards into the very heart of organisms.*

However, I promise to take a more active part and to interject with more telling questions in future if I feel that you are not being clear or that you are going beyond what I consider reasonable conjecture. I intend to be patient, for I know you want to take me on a hundred-year journey that has led to the wonders of biology in your own day. This is a journey I can take leisurely and with pleasure, for

I've got very little else to do, except listen to that incessant drumming of feet on the flagstones of the Abbey.

   Ever your most truly

                      Charles Darwin

THE RISE AND FALL OF
THE MOBILE P GENE

Dear Mr Darwin

I've enjoyed reading your response and the encouragement to continue with my efforts. I realize that I covered a lot of ground in my first letter that might have appeared both confusing and alarming. But I was keen to lay out my stall, so to speak, on some of the central issues that concern me. It is not every day that one can correspond with a scientist of your stature.

My starting objective in this and following letters is to explain the unexpected turmoil in the genetic material of all living things, whether virus, bacteria, plant, fungus or animal. An organism's full complement of genetic material is known as its genome. This genetic material is at the heart of every evolutionary process, and how it behaves or misbehaves, on a short or long timescale, is critical to the machinery of evolutionary change. This realization has been particularly strengthened over the past twenty years following on from some major discoveries that our genes do not necessarily behave in a regular and orderly manner, timelessly obeying the rules of inheritance as laid out by Mendel.

As I explained in my last letter, most genes, or rather the DNA of which they are composed, have a life of their own that is out of synchrony with the behaviour of the chromosomes. This is due to a bizarre range of DNA misbehaviour (or turnover, as we call it) that ensures, willy-nilly, that the segregation of DNA is decidedly non-Mendelian.

Although Mendel's rules result in the long-term stability of chromosome frequencies in a population (the Hardy–Weinberg equilibrium), this is generally not true for the genes, or for any of the important genetic elements that control the activity of genes during the development of an organism.

So the assumption in the neo-Darwinian synthesis that only natural selection can change a population's genetic constitution is not watertight. The non-Mendelian mechanisms of DNA turnover, no matter how slowly they operate, can do the same job. They can bring about a gradual 'modification by descent' (to use your telling phrase again) with the passing of the generations. I have called such a process 'molecular drive', an umbrella term covering a variety of different non-Mendelian mechanisms of inheritance. I'll explain what drives 'molecular drive' shortly with reference to the jumping P element.

It is critical, therefore, given the importance I attach to DNA turnover in 'molecular drive' that I explain one or two of these mechanisms in detail. Then there can be no confusion about their ubiquity and operation among both the genes and the all-important genetic elements that control when genes are 'on' or 'off'. The days have gone when such mechanisms could be marginalized as rare aberrations of DNA behaviour, as was the case on their first discovery. Only after detailed examination of all DNA misbehaviour can we make any realistic proposal about the origins of complex biological functions and adaptations. I hope to show you that there is an important role for selection in all of this, but not solely in the ways you originally proposed. Let me home in on a real example.

I have already given you a foretaste of one of the mechanisms that causes genes to move from one place in the genome to another, often to another chromosome altogether. This phenomenon was so startling when first proposed by Barbara McClintock in the 1940s, based as it was on some weird and unexpected genetic switching of colours in the kernels of corn, that it left most contemporary geneticists bemused and uncomprehending. As has been the case with several further DNA turnover mechanisms discovered after McClintock's jumping genes, these phenomena are grudgingly accepted but then largely ignored as nothing but background noise, with no real relevance to the evolution of adaptations. McClintock's genes were found to be misbehaving ten years before the discovery by Francis Crick and James Watson of the double helix structure of

DNA. Within two more decades, the same jumping phenomenon was observed in bacteria. This, in turn, led to an avalanche of similar discoveries from insects to humans, using the more precise experimental tools of molecular genetics. All genomes are riddled with the products of jumping DNA.

The P elements are coming

I'm going to base my first simple account of molecular drive, and the subsequent involvement of natural selection, on the mobile P element in the banana fruitfly *Drosophila melanogaster*, which has received worldwide attention in the past two decades.

Sometime in the early part of the twentieth century and somewhere in South America, a single small piece of DNA, measuring around 3,000 units of what are called nucleotide bases, found its way from one species of fruitfly (*Drosophila willistoni*) to another (*Drosophila melanogaster*). This piece of DNA is known as the P element. The precise means of infection is not yet known, but it is supposed to have involved a parasitic mite whose mouthparts acted as a hypodermic syringe as it repeatedly sucked out the juices of one fruitfly after another. While gorging itself on the flies and unbeknown to all and sundry, the mite accidentally transferred a single P element from *willistoni* to *melanogaster*. This one-off, improbable event transformed the life of the recipient species. Male and female gonads shrivelled up, chromosomes disintegrated and high numbers of mutant flies were produced. Unless some way of overcoming such mysterious and chaotic events could be rapidly found, this syndrome of effects would spell the end of the line for *D. melanogaster*, for the species would have suffered a catastrophic loss of Darwinian fitness: no gonads, no future.

We do not have any records of what precisely happened on that fateful night in South America and in succeeding generations of flies. But we do know from experiments in which P elements have been introduced from a strain of *D. melanogaster* flies carrying P elements to a strain of flies without them that all the symptoms I've described appear in the offspring. Collectively, the suite of disorders has been termed 'hybrid dysgenesis'.

Mr Darwin, I can already anticipate the first question on your lips.

If these mysterious P elements cause such havoc, how can strains of flies with P elements still be in existence? Surely they would have gone extinct, as is the way of the world for more than 99 per cent of all species that have ever inhabited the Earth? And how can one P element, inadvertently escaping from *D. willistoni*, infect complete populations of flies?

The answers lie in the fact that the P element is a mobile element. Once the first element managed to insert itself in the DNA of *D. melanogaster*, it could get down to the business of being replicated with the aid of the host. In fact, it turns out that there is a very clever way of doing this. First, the P element jumps out of the chromosome in which it resides using a special protein (enzyme) that I would like to call 'jumpase', but which is technically called 'transposase' for transposition. This enzyme is actually the product of the 3,000 or so nucleotide bases of the P-element sequence. But before I can continue with this story, I need to explain a little molecular biology: that is, the link between a gene and a protein.

A gene (a string of nucleotide bases that make up DNA) codes for a protein (a string of amino acids). There are only four types of nucleotide (abbreviated to A, T, G and C) and a gene can contain hundreds of units of each type. Each set of three nucleotides codes for a given amino acid, of which there are twenty different sorts. For example, triplet CGT codes for the amino acid arginine and AGT codes for the amino acid serine. The decoding of DNA into protein does not take place in the nucleus but in the rest of the cell surrounding the nucleus. Indeed, the chromosomes, and their DNA of course, never leave the nucleus. Decoding, therefore, requires an intermediate molecule that can act as a messenger between DNA and protein. This is the appropriately named messenger RNA. It consists of a single strand of nucleotides with the letters A, U, G and C, where U corresponds to T in the DNA. Messenger RNA uses one of the two strands of the DNA double helix as a template for its own synthesis by a process called transcription. So the order of bases in a gene is precisely transcribed into the same order of bases in the messenger RNA. The messenger RNA can then move out of the nucleus and into the surrounding cell where it is translated, by a complicated process that I won't describe, into a series of amino acids that define a given protein. Essentially, the unique series of bases that defines a gene specifies a unique series of amino acids of a given protein. And it is the specificity of amino acids in a protein that determines how it interacts with other molecules in the cell to bring about a given biological function.

So, back to our mobile P elements. The P element is an independent genetic unit, in that it has the ability to jump out of a chromosome. But where does the P element go to and what happens to the gap it has left behind? In reality, the P element can go where it wants to; it could get lost altogether by failing to reinsert itself elsewhere into the host's DNA, or it could survive by managing to find a suitable landing place for reinsertion. The gap could simply be closed up by the two loose ends of DNA joining up again through a 'joinase' enzyme (called DNA ligase). If this were all that happened, however, there would be no net increase in the number of P elements. One element would simply have moved from one place to another. But something rather bizarre happens. The gap is not closed up; instead it is filled with more P-element DNA. You might ask, so where did this come from? This gap-filling P element is made from scratch by using, as a template, the P element patiently waiting in place on the opposite chromosome. Remember, like all life forms above bacteria, fruitflies have two sets of chromosomes, one set from each parent. The double helix of P-element DNA on the alternative parental chromosome is unwound and copied, using a battery of host enzymes, in a way that allows the newly synthesized copy to fill in the gap in the opposite double helix from which a P element jumped. This remarkable process of gap filling is called gene conversion. It is essentially another non-Mendelian process with important consequences for evolution that I intend to describe in future letters, with your permission.

From the dual processes of jumping and reinserting, coupled to gene conversion, the P element can make more and more copies of itself and cause more and more damage. In a very short time, the exact length of which would depend on the frequency of jumping and replicating, an exponential growth of P elements would completely clog up the *Drosophila* genome. Such multiplicative growth, $2 \rightarrow 4 \rightarrow 8 \rightarrow 16 \rightarrow 32$ and so on, is the same process that worried Malthus in his essay on overpopulation – the essay that so influenced your own thinking on the root ecological cause of natural selection.

A simple case of molecular drive

The accumulation of P elements among the chromosomes of a single individual fly is the first requirement of the process of molecular drive. The second requirement is to explain how all individuals in a population acquire P elements. This is achieved unwittingly by sex, which is primarily a process for shuffling chromosomes. So let's imagine that a P element has been replicated by jumping from one chromosome to another. All chromosomes come in pairs and sex ensures that the two members of a pair are separated and enter into two different offspring. Accordingly, two new individuals at the next generation now have a P element, in each of which more P-element replication and jumping can occur. Once again, after another round of sex, the extra P-element copies enter yet more offspring at the third generation. Eventually, the P element will spread throughout all of a sexually reproducing population, with the passing of the generations. This is the molecular drive process: an outcome of P-element accumulation within an individual and sexual shuffling of chromosomes among individuals between generations. I've attempted to illustrate the process in a diagram that I hope you'll find useful.

Other DNA mechanisms can generate extra copies of genetic elements, such as unequal crossing-over, DNA slippage and gene conversion, to mention but three. All of these, like jumping genes, can lead to an internally driven spread of elements through a population of individuals. I won't go into these right now because I want to get back to the P-element story. In particular, it is important to examine the role of natural selection in all of this. You have probably realized that if the molecular-drive accumulation of P elements went unchecked, each fly would become so choked with P elements that it wouldn't be able to get off the ground!

Molecular drive and natural selection: a mutual accommodation

All over the world there are populations of *D. melanogaster* that are full of P elements, but not as full as one might expect were P element replication unconstrained. Most individual flies have 50–100 P

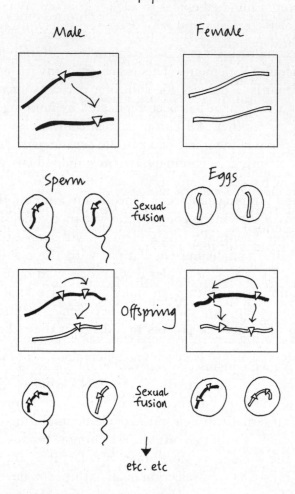

Spread of mobile element (Δ) through
sexual population

Jumping genes and sex: a simple case of molecular drive *One of the mechanisms that leads to molecular drive is the ability of genes and other genetic units to jump from one chromosome to another. The act of jumping often produces two copies of the gene in place of one. Chromosomes come in pairs, one from each parent, and each member of a pair enters a different sperm or egg. In the top left-hand corner I'm showing just one pair of chromosomes and just the sperm that inherit a chromosome with a mobile gene. Each sperm fertilizes an egg and so two new offspring can inherit the mobile gene, in each of which more gene copies are made as they jump from one chromosome to another. Eventually, through a combination of jumping between chromosomes and the shuffling of chromosomes by the sexual process, a mobile gene can spread through a population. This is called molecular drive. Other mechanisms can also spread genetic elements through populations in a way that is operationally distinct from natural selection. I'll describe these to you in later letters.*

elements. The same is true for about fifty other types of mobile element in *D. melanogaster*, each one of which is known to be represented by about a hundred copies in each individual. What controls the number of P elements and why don't flies succumb to the dire effects of hybrid dysgenesis?

The answer to both these questions is natural selection, but not in the way you might first suppose. It would be easy to imagine how flies suffering hybrid dysgenesis are gradually eliminated through their inability to reproduce. In that way, P elements are not passed on from one generation to another unless the rate of production of new P elements far exceeds the loss of individuals through natural selection. We could imagine a war going on between selfishly replicating mobile elements and the biological needs of the population to stay reproductively viable. Viewed in this way, mobile elements have been described as molecular parasites whose sole dream is to populate the world with their presence.

I disagree with this view. Such 'wars of the genes' are superficially attractive, but are not in keeping with what we know is happening in nature. Rather than eliminate individuals fast enough to prevent uncontrolled exponential growth of P elements, natural selection has been busy promoting genetic systems that can effectively shut down P-element jumping and so keep these elements out of harm's way. It is only when P elements jump that hybrid dysgenesis manifests itself, so locking elements into fixed chromosomal sites is an effective way of rescuing the reproductive capabilities of P-infected individuals. Essentially, natural selection has promoted the coevolution of systems of repression of jumping genes.

Hence, P elements, along with many thousands of other mobile genetic elements distributed throughout all forms of life – bacteria, plants and animals – can be tolerated as long as existing genetic processes in the genomes of the hosts can be adjusted, with the aid of selection, to coevolve with the mobile invaders and prevent any further damage. Natural selection has a very important role in improving the levels of internal tolerance, or accommodation, to mobile elements.

First come, first served

The evolution of repression of the P element is instructive for a variety of reasons. Foremost among them is the fact that serendipity has played a large role. The urgency with which a system of repression needs to be established is so great that selection has seemingly grabbed hold of any existing cell machinery to help in this endeavour. Accordingly, different repression systems became established in different parts of the world almost, but not quite, on a first-come, first-served basis. Not all the systems are as good as each other. Some are sufficiently powerful to remove any real signs of hybrid dysgenesis in a given population. Others are not as good and mild hybrid dysgenesis is constantly present under certain genetical and environmental circumstances.

Interestingly, one of the most powerful repression systems is transmitted only through the females. This indicates that repressor proteins need to accumulate in eggs to shut down the jumping of P elements in the next generation. Any proteins that may accumulate in sperm are not passed on because only the nucleus of the sperm enters the egg. Indeed, the repressor protein turns out to be a shortened version of the transposase – or jumpase – enzyme that I spoke about earlier.

By some coincidence of biology that we do not understand, the short form of the transposase can interfere with the true transposase and stop it making P elements jump. P elements can therefore produce, inadvertently, their own repressor. They know nothing about this, any more than selection does. That is the way the cookie crumbled. Nevertheless, faulty short transposase forms a repressor protein that rapidly accumulates in eggs and can be seized on by selection to overcome the debilitating effects of hybrid dysgenesis.

This type of repression system exists predominantly in North America and was the first to be discovered. P elements have now swept through all North American populations and the only populations left in existence without P elements are old laboratory stocks collected from nature before the 1940s. North American populations in nature manage to survive the presence of P elements because of the greater reproductive success of flies that accumulate repressor proteins.

In other parts of the world, selection has led to other systems of repression. They can be weaker in effect than the North American systems, but have the advantage of being passed through both sexes and not just through females. The European system is based on the fortuitous activity of complete P elements to produce much shorter copies of themselves through the removal of large sections of DNA within the elements. This has a twofold effect. The first is that such partially deleted elements can no longer make transposase at all, so the elements are no longer capable of jumping autonomously. If they want to jump, they need to rely on the production of transposase from other full-length elements. Second, of all the partially deleted P elements, there is one in particular that represses the jumping of autonomous elements. This is thought to be because the small RNA, or small protein, of this element binds to transposase and prevents it from carrying out its normal function. The transposase can be said to be effectively poisoned.

This particular small P element with repressor consequences was discovered in my laboratory. We called it the KP element because it was isolated from a population of flies originating in Krasnodar, Russia. Flies in most populations in Europe and Asia carry tens of copies of KP.

A simple case of molecular coevolution

So selection can promote the spread of any available naturally occurring system of repression. What's more, it can also aid in the promotion of an element like KP because the reproductive success of individual flies carrying KP elements is greatly improved. This promotion is in addition to the internal molecular drive process.

It is important to consider the response of selection to molecular drive if we want to have a comprehensive view of the evolution of a given biological system. The P-element story illustrates the types of interaction that can take place between non-Mendelian molecularly driven changes, often with detrimental effects on bodily functions, and natural selection. As I shall be illustrating further with other examples, selection's frequent solution to the spread of an adverse effect in a population is to promote internal tolerance to the culprit element rather than to throw out the individuals harbouring such elements. Selection is essentially promoting a 'molecular coevolution' between two genetic systems: in the case of P elements, between P elements promoting hybrid dysgenesis and fortuitously occurring systems of P-element repression. There is an internal, mutual accommodation of molecular systems, resulting from selection and molecular drive.

There are other important components of molecular coevolution which I shall leave to later correspondence. They revolve around important issues of biological redundancy (repetitions of genes and their controlling elements) and modularity. But enough is enough at this stage. I hope you are not too tired wading through this P-element story and its evolutionary implications. I have spared you some of the details of other systems of P-element repression.

This letter has exhausted me because it's all been done in old-fashioned handwriting, not unlike your own illegible notes, which continually plague your biographers. In your day, there was no choice, but today we have machines that print out each letter once one has pressed a given button. These machines are quite fanciful in allowing the user to move sentences and paragraphs around the text at will – a bit like mobile genetic elements, in fact. Unfortunately, I'm one of those people who can't think and press buttons (or chew gum) at the same time.

I would be thrilled to receive another letter of commentary from you. I know you thoroughly enjoyed corresponding with many scientists when you were younger and in this world, so I'm hoping that this enjoyable habit has not abandoned you altogether.

As ever

Westminster Abbey

Dear Dover

Your story of the single mobile P element creating havoc in whoever was infected by it brings me back to my own debilitations which plagued me all my life following my Beagle voyage. Could it have been some parasitic organism, picked up in South America, which multiplied inside me and was ultimately responsible for all the heart palpitations, agonizing stomach cramps, vomiting and rashes that laid me low and generally out of reach of my many scientific friends in London? Well, at least in my current out-of-body existence, all my mysterious illnesses are safely out of the way.

But now to more substantial matters. First, a matter of grammar. I can't help but notice that you use selection as a subject followed by an active verb: as in 'Selection ensures that ...', 'selection has promoted ...'. From the arguments you made in your first letter, you pointed out that (and I quote) 'selection is a one-off outcome of a unique set of relationships between a unique set of individuals and a unique set of ecological conditions.' I am prepared to accept some of the blame for 'natural selection' unavoidably sounding like 'artificial selection' as if some human agency is required to make the selections. It is not obvious to me, given that I would not wish you to change the name for my 'natural selection' mechanism, how you or anyone else can get around this language problem. It would be far too long-winded and a considerable mouthful were we all to engage in some type of phrasing that underlines the passive 'selective' outcome of the active interactions between individuals and their ecologies.

I recognize that it is far easier to propose that 'selection promotes the establishment of a genetic system to repress hybrid dysgenesis and P-element mobility' than to say that 'differences in the reproductive successes of individuals, contingent on the degree of hybrid dysgenesis each exhibits in a given environment, lead to the gradual selection and spread, over time, of the relevant genetic elements of repression through a population'. From the latter statement it is clear from the outset that it is individuals that get 'selected' (if we insist on using that term), given the luck of the draw

in the genes each has inherited, and that it is the genes that get
passively 'sorted'.

This is a nagging problem, of which I was well aware in my days.
It could be, from the persistence with which I see you emphasize the
point, that the pendulum really has swung too far in the direction of
natural selection as an active process. That said, I suspect that there
might be other reasons why you emphasize this point that you are
keeping back from me. Maybe you'll wish to expand on this in
future correspondence.

With respect to the P-element story, it is amazing for me to hear of
mobile genetic elements. Not only have I only just heard about
well-behaved Mendelian genes, but now you tell me that genes can
be seen flying all over the place, even crossing species barriers. I'll
certainly have to take your word for it, not that I doubt your scientific honesty. Maybe you could supply me with some reference
sources for the main events you describe.

You have emphasized that the role of natural selection is more to
do with the promotion of genetic systems of repression to prevent
hybrid dysgenesis than with the elimination of the particular individuals carrying too many P elements. How do you know, however,
that elimination has not happened repeatedly? In other words,
might there not have been many independent invasions of P elements from D. willistoni to D. melanogaster, and might many such
populations of D. melanogaster have gone extinct, leaving just the
one invaded population to survive – the one population fortunate
enough to evolve a repression system? After all, selection is not
overly concerned with its biological consequences. It really does not
care about extinction; or, to put all of this into the current passive
mode, extinction could very well be the natural order of the day,
just as it has been throughout evolution, and the final outcome for
most invasions has been the total absence of individuals carrying P
elements.

Finally, if I may continue for a little longer, although I can see
that P elements are fascinating for you and, in turn, you have
aroused my interest in such mobile genetic elements, I cannot see
any direct link between them and the basic biological adaptations
such as eyes, wings and beaks that featured so largely in my evolution books. Mobile genes might have a separate evolutionary tract

distinct from Mendelian genes, but are they relevant to anything of real interest in biology? Just curious.

Your obedient servant

Darwin

WHEN IS AN ADAPTATION NOT AN ADAPTATION?

My dear Darwin

As I was expecting, your last question in your most recent letter hits the nail on the head, or, as we say over here in our late-twentieth-century western world, it asks the $64,000 question. This is the size of the ultimate prize in a game in which contestants answer quite simple questions as part of mass audience entertainment. This prize would have been more than sufficient to buy your country house in Kent.

The evolution of bat-eared foxes by means of television

The question of the origin of adaptations is not a simple one to answer. First of all, as I explained in earlier correspondence, your process of natural selection and its final product have been given the same name: adaptation. This is unfortunate in that cause and effect are terminologically locked together, leading to the easy belief, prevalent in many quarters, that if a given function can be described as an adaptation (for example, wings for flying) then it must have arisen through natural selection. The proof of the pudding of selection is taken to be in the description of the function as an adaptation. With few exceptions there is no concern for the real weakness of natural selection theory, which is its ability to account for too much. Sir Peter Medawar, an eminent man of science and British Nobel laureate, expressed this succinctly when he reflected that 'natural selection has such an enormous experimental facility that one could hardly imagine anything it could not explain. Now the danger of this is that it rules out any incentive to enquire about any other possible mechanism that could explain the observed facts.'

The reason for the universal acceptance of natural selection as a mechanism for the origin of diverse forms of life is that it is based on the simple fact that some but not all diverse life forms have survived. That some forms of life exist and continue to survive in particular environments whereas others have not is incontestable. If the evolutionary process of adapting to particular environments is defined as one of differential survival, and the mechanism responsible for this process is defined similarly by differential survival, then no further independent proof is required of the system other than that given by the formal definitions. Those that survived are by definition better adapted to the environmental circumstances that ensure their survival. So differential survival, measured in terms of relative number of offspring populating the next generation, becomes synonymous with the concept of adaptation. This being so, we are left with an explanation of the diversity of life as a consequence of an ever-increasing radiation of life forms adapted to new environments under the aegis of natural selection. Everything is for the best in this best of all possible worlds, and all existing components of form and behaviour have their adaptive utility. Survival and adaptation are often merged in a warm fog of wishful and more-or-less tacit equivalence.

> That was a way of putting it – not very satisfactory
> A periphrastic study in a worn-out poetical fashion
> > T.S. Eliot, *The Four Quartets*

Despite the fact that this problem of tautology is an old and hoary chestnut, it has not been satisfactorily answered. We can go round and round the problem but never quite break out of the circle, although I attempted to do as much in my very first letter. Indeed, the ease with which we take refuge in the circle is endearing. I cannot resist giving one homely example to drive the point home.

Recently a British television award was given for the best wildlife film of the year: a documentary on the seasonal change in environment and life around a single water-hole in the Namibian desert. We are entertained by life in the raw. No sooner are our sentiments aroused by the lovable antics of one animal than it disappears, whole or in part, down the gullet of another. Nature red in tooth and claw under the strong Namibian sun. All appealed to the magic of natural selection as sole provider of the adaptive wonders required for survival in the unfriendly and inhospitable surroundings. One such wonder lies in the shape of the abnormally large ears of the bat-eared fox. This creature is of normal fox proportions except for its bat-like ears. The voice-over blandly proffered the explanation that such extremities have evolved by natural selection for greater efficiency in aural tracking of subterranean prey. The bat-eared fox, accepting its cue, obligingly inclines its ears to the ground. There could be no doubt that such ears are of great utility in this pursuit, just as there could be no doubt in the mind of the commentator as to the reason for their evolution. The present-day use of these particular ears is taken as *prima facie* evidence for their evolution as an adaptive device that increased the survival of past generations of animals that possessed them. The force of Sir Peter's warnings is no more apt than in this commonplace extrapolation from current usage to past evolutionary change. Why are professional biologists willing to accept this line of reasoning?

In my opinion the answer lies in a mixture of historical precedence (natural selection came first) and the psychology of a preference to accept simple rather than complex matters. But bat-ears, as used today, are as explicable as adoptations or exaptations; as well as adaptations; or, more realistically, a mixture of all three.

Exaptations and adoptations: a brief introduction

'Exaptation' is a new word to describe biological functions that arose by a variety of other means. For example, they could arise by a widespread genetic process called 'neutral genetic drift'. 'Modification by descent' (a change in the genetic composition of a population) can arise through accidents of genetic sampling and not because of the relative reproductive successes of individuals carrying the relevant gene.

For instance, an individual male produces many more sperm than are actually used. It is possible that the genetic composition of used sperm is not the same as the bulk sperm, simply through accidents of sampling. Different versions of a particular gene are known as alleles. If the bulk sperm is composed of, say, 50 per cent carrying gene *A* and 50 per cent carrying an alternative form of the gene (or allele) *a*, the sperm actually used for fertilizing the eggs could have carried more of one allele than the other. When sampling errors occur repeatedly, it is possible that one allele ultimately spreads in a population at the expense of another. The process is called neutral genetic drift because the genes involved are neutral with respect to reproductive success, and natural selection can be said not to care one way or another which allele increases or which allele decreases in frequency.

Importantly, once a given neutral gene has spread by accident, often with a corresponding neutral change in phenotype (the observable characteristics of an organism), it is possible that it might acquire a new function that could affect reproductive success. This could arise, for example, with a later change in the environment. But if the original gene had spread throughout a population by neutral drift, then most or all individuals will be affected concurrently.

So we have here a later functional co-option of shared genetic novelties that were originally neutral. In such circumstances, it is not possible to claim that the ensuing biological functions affecting reproductive success are adaptations, for this would imply that they arose from the beginning by the grinding process of natural selection. Two well-known evolutionary biologists, Stephen Jay Gould and Elizabeth Vrba, decided that a new word was needed, and they coined the word 'exaptation'. There are other ways in which exaptations might arise, but my favourite method is as I've just described. There is a clear operational distinction between the establishment of an adaptation and the establishment of an exaptation.

Similarly, novel biological functions may arise through molecular drive. I have explained one of the mechanisms leading to molecular drive in my letter on mobile genes, and I have more to say about novel functions arising through molecular drive in due course. Let us accept for the moment that the internal non-Mendelian mechanisms of turnover can bring about a modification by descent in both genotypes (the genetic make-up) and phenotypes that is operationally distinct from both natural selection and neutral drift. It would be confusing and inappropriate to name these new functions as either adaptations or exaptations. Accordingly, I have called them adoptations. This word is only one mutational letter away from adaptation, so it shouldn't cause too much disturbance. In fact, the word is chosen specifically because a molecularly driven change in a population (by, for example, 'gene conversion' or 'jumping genes') allows the population, on average, to *adopt* a component of its existing environment not previously accessible to it. Mr Darwin, it is not possible to explain in this letter the dynamics of population change under molecular drive that allow me to come to this conclusion. I will need to give a proper account of this at a later stage, once I've finished describing the relevant non-Mendelian mechanisms of DNA turnover.

What I can say here, and to come back to your question, is that I believe that all biological functions (no matter in what organism) have arisen through intimate and complex contacts between all three major mechanisms of modification by descent: natural selection, neutral drift and molecular drive. So much so that every novel function can be viewed as an interesting mix of all three products: adaptations, exaptations and adoptations. Indeed, I would go so far as to say that in our current state of knowledge of biology we are not in any position to quantify the relative contributions of any of these three processes and their products, for any given function, especially when we are trying to explain the evolution of eyes, wings, giraffe necks, brains and the panoply of functional wonders that lie before us.

We are beginning to acquire the experimental tools to tease out the contribution of one or other process in one or other function, and I shall provide some interesting examples in due course. But we are nowhere near relieving our deepest ignorance about the biological world around us. This offers an exhilarating prospect for the future, but cannot provide easy answers for the moment.

So, to return to your question about mobile elements and adaptations. Mobile elements can interfere with the genetic material in a variety of ways; it all depends where they land. If they land inside genes or near to the genetic control circuits that decide when a gene is turned on or off during development, they could affect biological function, whether these are adaptations, exaptations or adoptations. When I get down to telling you about the genes that affect the form and behaviour of animals, and how these genes are controlled during development, I'll be able to give you concrete examples of where genetic mobility and other mechanisms of turnover affect all such operations.

Honing locks and keys

My purpose is not to go into each and every known effect of mobile elements on the natural and proper workings of the genes in all forms of life. This would become an exhaustive catalogue that even you, Mr Darwin, reposed in your timeless state, would tire of. Indeed, I rather self-consciously wrote the words 'natural' and 'proper' in the preceding sentence given the implicit and almost unavoidable assumption that there is something in biology that is 'natural' and 'proper' waiting to become 'unnatural' or 'improper' on disturbance by a mobile element, or through any other form of genetic rearrangement. My hesitancy is born of the fact that we have many convincing examples in which the genetic architecture of sequences surrounding particular sets of genes is simply the current endpoint of many accidental events resulting from the non-Mendelian systems of DNA turnover, of which mobility of elements is just the tip of the iceberg. There is nothing 'natural' in how biology works today, any more than there was throughout its nearly four billion years of evolution.

To some extent your theory of natural selection, especially in its modern guise, has always recognized this accidental component. Straightforward mutations of gene sequences occur as one-off, accidental events in a single individual without any regard for future usefulness in the reproductive prowess of the individual. Mutation has always been viewed as blind to any adaptive needs. Adaptation is a result of natural selection, not a result of mutations preordained to fulfil some useful function.

An image of a unique lock and its unique key neatly encapsulates the process of natural selection, both externally in the ecology and internally within our bodies. The external lock is the ecological niche to which only the appropriate key (organism) is found by natural selection, until the day an improved key arises and takes its rightful place. Always, however, one unique key for one unique lock. Similarly, two molecular structures involved in a given function represent an internal, mutually beneficial, lock-and-key relationship continuously honed and improved by natural selection. The physical relationship between an enzyme and its substrate (the substance it acts on) is a case in point. Adaptations, by definition, are the products of a continuous and unavoidable process of refinement. There is little room for flexibility and tolerance, other than in the most trivial of ways. What is 'proper' and 'natural' (to use my earlier words) is what has been permitted to survive and reproduce by natural selection. All other functions are assumed to be evolutionary dead-ends: unnatural and inefficient monsters in the outer swamps surrounding the natural tree of life. To use modern-day parlance, our current biological functions are often considered to be the lean machines generated by our mean genes in their endless goal of selfish reproduction.

Just-so storytelling

Why are so many evolutionary biologists convinced of the unshakeable truth of the overarching relationship between novel biological functions and their origins as adaptations produced by natural selection? The fault line goes back to Mendel. As I wrote earlier, his rules of inheritance led to the realization that without natural selection there could be no change in the frequencies of different versions of a gene in populations. As natural selection was the only known process in the 1930s that could promote the spread of novel genes and produce ever more refined adaptations, it is not surprising that all functional novelties are believed to be circumscribed by this process. We might not actually know what happened during the evolutionary history of the giraffe's neck, but until recently we have had no cause to question that it arose by natural selection, for natural selection is all that there was.

All we need do, if we are so inclined, is to tell the most plausible 'just-so' stories about the adaptive significance of ever-longer necks in the lineage leading to giraffes. For many years, longer necks were considered to have evolved in response to the urgency of reaching leaves high on the trees. More recently, the story has changed to one in which the adaptive superiority of longer necks is in response to sexual prowess between sparring males. No matter what the story of the decade might be, our school textbooks are certain in their description of how the giraffe got its long neck – solely through the process of natural selection.

We are selected, therefore we are

This problem of just-so story telling is not some minor irritation to do with the perennial problem of giraffes, dismissable as some naive caricature of what you really proposed in your theory of evolution. The problem runs much deeper and wider, embracing many new disciplines of evolutionary psychology, Darwinian medicine, linguistics, biological ethics and sociobiology. Here quite vulgar explanations are offered, based on the crudest applications of selection theory, of why we humans are the way we are. There seems no aspect of our psychological make-up that does not receive its supposed evolutionary explanation from the sorts of things our selfish genes forced us to do 200,000 to 500,000 years ago.

Did you know that women are genetically programmed to read maps badly? That step-children suffer more than usual at the hands of step-parents because of past genetic imperatives to look after one's own genes only? That 'survival of the prettiest' is a seriously proposed evolutionary adaptive process, as if ugly people do not mate and reproduce? That we are mentally wired-up and doomed ruthlessly to compete – in particular, men. All of these and many more are accepted as naturally selected adaptations. And all currently diminish free will and choice.

Not only is there the embarrassing spectacle of psychologists, philosophers and linguists rushing down the road of selfish genetic determinism, but we are also shackled with their self-imposed justification in giving 'scientific' respectability to complex behavioural phenomena in humans which we simply do not so far have the scientific tools and methodologies to investigate. There is a naivety about genetic determinism in both evolution and development that signifies intellectual laziness at best and shameless ignorance at worst when confronted with issues of massive complexity.

A solution looking for problems

Once I have explained how genes affect development and behaviour, I shall attempt to explain how we might begin to dissect the evolution of the giraffe's neck (or any other aspect of a living organism including our human nature) to tease out the diverse processes of evolutionary change, including natural selection, that went into shaping the giraffe. As Peter Medawar said, there is an unfortunate sense in which your natural selection theory is too powerful for its own good. It has become the Swiss Army knife of biology: an all-purpose solution looking for problems. I hope to show you that for many scientists the 'problems' of biology are so ill-understood and so ill-defined that they are not at all sure what it is they are asking selection to 'solve'.

As ever

Gabriel

Westminster Abbey

Dear Dover

Thank you for your last letter, which I'm afraid to say leaves me at times more confused than enlightened. It seems to me that there are a lot of ideological battles in your modern world that colour the simple scientific description of the factual base on which we can mould a theory of evolution. I'm not naive enough to think that in my day it was all so simple, that all I had to do was to write On The Origin of Species *and the world would fall at my feet. Notwithstanding that the first printing sold out in the first few days – not unexpected given the brazen title I chose – my theory of natural selection was accepted by only a few reputable biologists in England, Germany and America. I've always wondered where those first few thousand copies went to – probably bought by overzealous clerics to ensure that they were put under lock and key and left unread by the flock!*

As you well know, we had our ideological battles too, both inside and outside biology and not entirely dissimilar from yours. We had a true royal battle going on with the religious establishment and a major disinclination of all shades of biologists to accept my idea about how small, inherited differences between individuals could be the thin end of the wedge that separates species. Although I was heavily criticized for saying very little about the origin of species, particularly given my book's title, I've always believed this criticism to be unfair. After all, I did supply, along with Wallace, a perfectly plausible general mechanism for the gradual widening over time of genetic differences between organisms. I could not point to one solid example by way of proof that, yes, such an observed difference between two species arose by natural selection. But this was not the point. My theory was designed to make people think of how any such differences might have arisen, given the known facts about heritable variation between individuals and the need to survive and reproduce better than their natural competitors in an overpopulated environment. This conjecture of mine, simple enough in its outline, caused a major furore in the biological world. I agree with you that the Mendelian rules of inheritance, while giving a very welcome boost to selection theory, are not required for an understanding of the mechanism of selection. Despite this, biologists took an inordinate amount of time to come to terms with my theory, and not always because of the strength of their religious beliefs. There is clearly something about evolutionary biology that makes it prone to controversy. Some of the points you raise suggest to me that there has been no serious lessening of the polemics – I'm not sure that physics and chemistry suffer to the same degree. Can you put your finger on what the problem might be? Maybe you could write back on this curious aspect of evolutionary biology.

But before you do that, you might think it appropriate to clear up once and for all this business about selfish genes. You have mentioned them several times in your remarks about the mechanism of evolution. It is not clear to me what the original idea might have been and why you are against it. I feel that unless we get this matter out of the way, there will be an unresolved polemical battle running through the discourses that could get in the way of the things that I would really like to hear about. An alternative tack would be just to ignore the 'selfish gene', but from what you've already said about its all-pervasiveness in many other disciplines of human psychology and social behaviour, it might not be advisable to ignore the problem and hope it will go away.

We had similar problems in our day with Lamarckian evolution – the belief in the inheritance of characteristics through their use or disuse in the lifetime of individual organisms. Although it was clear to me that my theory of natural selection was an entirely different idea and could not genuinely accommodate the inheritance of acquired characteristics, given the obvious contrast between evolution by chance and evolution for a purpose, I became ambiguous about this issue towards the end of my life, and did not stand firm enough. My excuse now in the light of what you've told me is that we did not appreciate Mendel and the way in which genes are transmitted between generations; nor did we know how genes contribute to the development of a multicellular organism from the first fertilized egg cell. I look forward to what you are going to tell me about this.

I'm not sure whether the selfish-gene concept is a similar problem for you and your colleagues, immersed as you are in the molecular nature of genes and in the intimate processes that govern development. Does the selfish gene interfere with your emphasis on the role of the individual in the selection process? You claim that in bringing back the individual to centre stage you are returning the focus of evolution to where I had it, and from your description of natural selection as the passive, one-off outcome of individual interactions with the environment, I wouldn't disagree. Hence I'm perplexed by the selfish gene: is it real or not? Should I be concerned about it?

Well, Dover, I've given you a lot to think about. I don't want to dictate the course of your account of the new discoveries in evolutionary theory and biology in general, but, if I might be so bold, I could be in a similar position to other biologists not entirely familiar with your modern-day developments and would find it easier if you tried to clear up one problem at a time. Selfish genes: in or out?

Thank you for your patience.

Charles Darwin

THE IGNORANT GENE

Dear Darwin

I still can't bring myself to call you Charles, but maybe I'll get around to taking the plunge shortly, with your permission.

Evolution as ideology

I'm taking your advice to rid myself of the nagging problem of the 'selfish gene'. My hope is to clarify the problem for you, as I see it, and through such criticism to begin to build an alternative view of biological evolution. There is a deep sense in which the idea of the selfish gene has so misused and usurped your theory of natural selection, and is so relentlessly pursued by its originator, Richard Dawkins, that biologists and non-biologists alike, not always primarily engaged with evolution theory, are being seriously misled.

As I briefly mentioned in my first letter, the belief that there is a specific gene or set of genes *for* every observable characteristic of an organism (eyes, language, sexuality and so on), and that the current ubiquity of these genes is due to their own selfish pursuits in times past, is currently setting the intellectual agenda for many important aspects of human nature and society. I believe that we have a major intellectual crisis on our hands. Notwithstanding your own well-known modesty, you would be shocked at the way in which you have become the spiritual head of a new ideology. Alas, poor you!

Anti-Dawkins

I want primarily to present in this letter a simple anti-Dawkins description of the inside-out world of selfish genes and the fundamental ignorance of the workings of natural selection that it reveals. There is an historical precedent for the anti-mode of expression. You might recall receiving a copy of the first volume of a large German book, *Das Kapital*, by a Mr Marx, who proclaimed to have read and admired your evolution theory. It is said that this remained largely unread by you, which is a pity because there are some interesting parallels between his theory of the causes of human historical changes and your own theory. But I shall leave this for the moment. My point about Mr Marx is to introduce his lifelong close collaborator and popularizer, a gentleman by the name of Mr Engels. It was Mr Engels who in 1878 wrote a brilliantly mischievous book called *Anti-Dühring*, in which Herr Eugen Dühring's self-proclaimed 'revolution in philosophy, political economy and society' was systematically exposed for what it was. From the pieces of this exposure, Engels managed to achieve two things, one intended, the other unintended. The first was a constructive account of an alternative analysis of history and society, and the second was the immortalization of Herr Dühring, who would otherwise have gone the way of many another nineteenth-century European systems-builder.

There are notable similarities between Dawkins and Dühring, and, like Engels, I intend to use a provocative anti-Dawkins polemic as a device to paint a much larger picture of some of the late-twentieth-century excitements of new genetic discoveries and their evolutionary implications. These developments have no room for Dawkins' misappropriation of your theory of natural selection, as embodied in the selfish-gene illusion. Dawkins and his closest followers are striving to become the village schoolmasters of the global village.

> While words of learned length and thundering sound
> Amaz'd the gazing rustics rang'd around,
> And still they gaz'd, and still the wonder grew,
> That one small head could carry all he knew
>
> Oliver Goldsmith, 'The Deserted Village'

One step forward, two steps back

Dawkins claims that he was appalled, back in the 1970s, by the prevalent sloppiness of animal behaviourists in explaining natural selection on the basis that what was selected was 'for the good of the species' or 'for the good of the group'. Dawkins' discomfort with the supposed misappreciation of the mechanism of natural selection in his own discipline is not sufficient justification, however, for substituting his mistaken notion that natural selection is 'for the good of the gene'. The gene, and not the species, group or individual, becomes the unit of selection. The only serious attempt to accommodate the gene as the unit of selection concerns the phenomenon of altruism, as if this behavioural trait, considered by sociobiologists to be the last remaining unsolved problem in evolution, is living proof of the gene as the unit of selection. Altruism is when an organism helps another unrelated individual survive, to the detriment of its own Darwinian reproductive success (or 'fitness'). This is not supposed to happen if genes control the individual for their own selfish gains.

But in the wider context of the living world of viruses, microbes, plants, fungi and the majority of animals, this phenomenon is marginal in the extreme and can be explained by selection operating at the level of groups of individuals. I'm sending you a recently published book by Elliott Sober and David Wilson, *Unto Others: The Evolution and Psychology of Unselfish Behaviour*, which explains convincingly how altruism can evolve by group selection. In any event, Dawkins is pursuing a global explanation for the evolution of the panoply of complex adaptations in all life forms wholly in terms of 'for the good of the genes'. Dawkins' selfish-gene concept stands or falls on the validity of the proposition that natural selection works at the level of genes in their relentless drive for their own self-propagation.

Where did Dawkins go wrong?

To justify the concept of the gene as a unit of selection, Dawkins makes several assumptions, all of which are wrong.

The first assumption states that what is important in evolution is the survival of a unit that must be an immortal, self-replicating entity, and the gene (and not the organism) is the only true immortal, self-replicating entity. From this basis Dawkins concludes that 'the fact that genes in any one generation inhabit individual bodies can almost be forgotten'. His logic leads him to state that 'elephant DNA is a gigantic program which says "Duplicate Me" by the roundabout route of building the elephant first'. The elephant is a digression. Within the bounds of this logic, Dawkins' metaphor for evolution as a river flowing out of Eden is apt: individuals represent transient and stationary collections of genes that ultimately disperse and regroup in future downstream individuals owing to the sexual shuffling of genes between generations. Only the genes are eternally flowing and self-replicating; only the genes are the units of selection, for they directly control which of their robotic digressions go forth and multiply.

The second assumption concerns the nature of evolved biological structures and functions. For Dawkins, all structures and functions are adaptations, and all are 'improbable perfections' that could have arisen only by natural selection. He therefore writes in his book *Climbing Mount Improbable* that 'eyes are perfections of engineering – should any parts be rearranged then that would make them worse'. This view leads naturally to his opinion that 'each species is an island of workability set in a vast sea of conceivable arrangements most of which would, if they ever came into existence, die'. Organisms become tiny islands surrounded by an 'ocean of dead unworkability'. In this scheme of things, we need to beware of the unnatural monsters lurking in the swamps beyond the natural tree of life – naturally born of natural selection.

The 'paradox of the organism'

There is from the outset an inherent tension, recognized by Dawkins, between these two assumptions – the first that the gene, in its self-replicating eternity, is the single, selfish unit of selection, and the second that complex biological structures and functions are 'improbable perfections' of natural engineering, made possible only through natural selection. If the dream of each and every gene is only to make more copies of itself, come hell or high water, then how can complex parts of individuals (eyes, hearts, photosynthesis, consciousness, language) come into being, requiring, as they do, an intimate, coordinated behaviour of tens of thousands of similar dreamers?

Dawkins' 'paradox of the organism' arises painlessly from his belief that the organism should by rights 'be torn apart by its competing replicators'. He is so convinced of this paradox that he drives himself, without too much modesty, inexorably to the ultimate conceit: 'the organism functions as such a convincingly unified whole that biologists in general have not seen that there is a paradox at all!'.

Yes, we have no paradox

Dawkins' 'solution' to this 'paradox' is that genes subdue their individual selfish pursuits by collectively deciding on a shared list of desiderata (Dawkins' coinage) which ensures that they all end up in bodies producing successfully functioning sex cells. To quote, 'They all "agree" over what is the optimum state of every aspect of the phenotype, all agree on the correct wing length, leg colour, clutch size, growth rate, and so on.'

At this point, I think it is legitimate to ask whether the genes are selfish or not. Is each and every gene a unit of selection or is the organism (the collective love-in of happy, hippy desiderata) the unit of selection? Technically, to exit from the cul-de-sac of his 'paradox', Dawkins has, consciously or not, been forced to redefine the gene in terms of transient and ephemeral structures and functions of the phenotype, which, as I argued earlier, are the only true determinants of selection. By phenotype I mean everything beyond the gene, from the protein products of the genes through the network of molecular interactions to the all singing, all dancing, reproducing individual organism. For Dawkins, 'surviving genes are those that flourish when rubbing shoulders with successive samplings from the genes of the whole species, and this means genes that are good at co-operating.' So the cooperative gene–gene *interaction* now becomes the inevitable focus of Dawkins' world-view, and the interaction *is*, of course, the phenotype. The gene-as-organism has become Dawkins' roundabout route to retaining his selfish gene while solving the pseudo-paradox of the organism – a case of having his cake and eating it.

Dawkins' selfish genery is genetically misconceived, operationally incoherent and seductively dangerous. There has to be a way out of this impasse if we are to return the genes to their rightful place and uncover the wonders of their true roles in biology. The genes need to be released from the charge of unreconstructed hooligans flashing their absurd desiderata lists.

More soberly and recently, Dawkins has substituted the desiderata image with the bizarrely contradictory phrase 'selfish co-operator'. This is yet another linguistic device that he has designed to rescue himself from his problem of the 'paradox of the organism'. By this means Dawkins insists that the unit of selection is still not the uniquely constructed individual organism but an average set of genes, drawn from an ever-swirling gene pool, who have collectively realized that it is in their best selfish interests to cooperate in building an organism. But the molecular acts of cooperation *are* the organism; and it is the unique set of interactions that go to make up an individual, not the abstraction of an average set, that remains the determinant of evolution.

The introduction of the 'selfish cooperator' phrase indicates that Dawkins is half way to recognizing that gene cooperation and molecular coevolution are the name of the game, and have been so since life's origin. The phenotypic product of internal cooperation and coevolution, with its ability to reproduce, sets the evolutionary agenda and nothing else. All that is left is for Dawkins to drop the epithet 'selfish', not because of its emotive overtones, but because of the falsity of his claim that such a gene is an autonomous, self-replicating unit of selection. Saying 'I was wrong' is a way for scientists to gain prestige among their peers as Dawkins has pointed out in his book *Unweaving the Rainbow*. But, dear Charles, I'm not holding my breath over this.

Who reproduces?

There is no such thing as a free-standing, self-replicating molecule in biology, and there probably never has been. DNA relies for its replication on tens (maybe hundreds) of protein enzymes. In turn, the correct synthesis of proteins relies on the coded information in the sequence of four nucleotide bases of DNA and on the many enzymes involved in translation. In all probability, there has been an intimate coexistence and coevolution of the different kinds of macromolecule (DNA, RNA and proteins) in the soup from the beginning. And it is this mixture, making up the first prototype cells, that would have had the capacity to reproduce. The more recent finding that some types of RNA molecule can act as enzymes is not evidence that they are self-replicating. Not one free-standing self-replicating RNA molecule has emerged from quadrillions of randomly produced RNA molecules in laboratories around the world. The most that has emerged so far is an RNA enzyme that can put together a string of six nucleotides, if a suitable template is supplied, and no more. The existence in all contemporary life forms of key enzymes and other functional moieties that are mixtures of short stretches of RNA and amino-acids testify to the intimate cooperative associations that were probably in existence since the beginning.

The only free-standing biological entity capable of self-reproduction is the cell – the full-blown phenotype, part of whose diversity is governed by the genotype. The genotype cannot self-replicate; the phenotype can. For sexual organisms, whether single cells or multi-cellular, we need to add the caveat that reproduction, by the very nature of the sexual process, is not exact. Nor does it involve only one side – it takes two to tango. But the resulting offspring, although unlike either parent, are obviously more like their parents than they are like any other parents. On this basis, sexual organisms can be said to reproduce autonomously in pairs; genes cannot reproduce either singly or in pairs.

Selection requires a living entity to reproduce (or overproduce, if we return to your original reliance on the Malthusian principle). This is achieved by the organism. Selection also requires the inheritance of coded information from one cell to another or from one generation to another. This is achieved, to all intents and purposes, by the DNA. Dawkins' first major blunder is to suppose that, because there is a genetic contribution to the form, behaviour and reproduction of the organism, and because this contribution is inherited, the gene is the unit on which selection acts. The second big blunder is to suppose that, because there is a genetic contribution to a given complex function, natural selection, acting alone, is the only mechanism capable of achieving the 'improbable'.

As I have emphasized in my earlier letters, natural selection is nothing more than the one-off, passive outcome of a unique set of interactions of newly established, unique individuals with the local environment, in each generation. So whatever phenotypic selection and genetic sorting might have happened in any given generation, the same choices and sortings will never reoccur. On this basis, natural selection is not an active process (that is, there is no force directly selecting phenotypes or selecting selfish genes); indeed it is not a process, as such. You yourself advised against the potential confusion over this issue – advice ignored by generations of so-called Darwinists. It is illegitimate to give 'powers' to individual genes, as Dawkins would have it, to control the outcome of selection.

Long live the ephemeral phenotype!

It would be simpler and more logical to abandon completely the idea of the gene as the unit of selection, and to focus instead on the role of the ephemeral phenotype: that is, the all-purpose, self-reproducing product of motley genetic interactions, unfolding in a given environment. There are no genes for interactions, as such: rather, each unique set of inherited genes contributes interactively to one unique phenotype. The transient nature of unique individuals, coupled to their ability to reproduce, therefore makes them the true determinants of selection. Selection is here and now, not everywhere and for all time.

Selection can be pushed any which way, at any generation: we cannot make predictions from what happened in one generation about what might happen in the next. This is because of the sheer variety of unique phenotypes generated by sex at every turn. As you pointed out to me in your very first letter, a shifting population of organisms is totally different in structure from a constant population of water molecules that can be predicted to behave in ways prescribed by known constant physical laws.

This is not to suggest that a trend towards, say, longer necks of giraffes was ever a one-off, passive outcome of differences in reproductive success of phenotypes in a single generation. It is perfectly logical to envisage, as you and others have explicitly proposed, that the longer-necked phenotypes of any given generation are the result of the more reproductively successful longer-necked phenotypes of the previous generation, who passed on the genes that contributed to longer necks. The outcome of natural selection, even as a passive, one-off comparison of reproductive successes among unique individuals in any given generation, depends on the passive, one-off comparisons among individuals of the previous generation, and the one before that and the one before that...

This line of apparent regression should not fool us, however, into substituting the workings of selection with imaginary genetic units of selection, eternally self-replicating with their fixed selective 'powers'.

No genetic blueprints

The form and behaviour of a developing individual in any given species is not directly specified in the DNA. I shall shortly explain to you that there is no blueprint, or set of instructions, in the DNA saying 'make me an elephant – I feel like digressing'. Genes are wholly 'ignorant' of their effects on development. The final, multicellular individual, bristling with your Darwinian adaptations and much else, is nothing more than the product of piecemeal interactions of proteins with proteins, proteins with DNA, proteins with RNA, and RNA with DNA. The interactions are local and blind to everything that has gone on before and everything that has yet to come during development. Development is a consequence of a progressive unfolding of gene activity in particular cells and at particular times. Which genes are switched on or off is under the control of the protein products of other genes. So the specific pattern of unfolding that ensures that, for example, each human fertilized egg develops, on average, into a multicellular, reproducing human adult is intimately bound up with the evolutionary history of the genes and their regulators. As with everything else, these have been sorted as after-effects of millions of differential successes in reproduction among unique and ephemeral phenotypes, involved with their unique environmental conditions, stretching back in time to the inception of the first DNA–protein coevolving entity. They have also been strongly influenced by millions of instances of molecular drive and neutral genetic drift.

Unoccupied space: natural or unnatural?

As I explained in my previous letter, one prevalent image of natural selection is that it provides a solution (a key) to a given problem (a lock). The underlying assumption of selection is that there is always sufficient genetic variation for the current 'adaptive solutions' to inherit the Earth. What's more, it is assumed that all biological functions are naturally engineered by natural selection. No other functional and reproducing entities, within the assumed strictures of locks and keys, could have successfully evolved.

If we were to draw a three-dimensional graph and place within it the shapes of all known organisms, very little of the space would be occupied. Small clusters of occupancy representing similarly shaped species would be seen, but by and large the space would be empty. Why is this? Why are there no known shapes of either living or fossil species that would occupy other parts of the three-dimensional space of all possible forms?

For Dawkins, only the unnatural monsters of legend lurk in the largely unoccupied space of phenotypes. Dawkins once likened this outer darkness of unoccupied space to rows upon rows of empty shelves in his example of a multidimensional museum of all possible shapes of snail shells. For him, this vast unoccupancy of shelves is due to unsuccessful mutant shapes that were of no use to selection in its quest to produce robotic miracles of biological engineering, with their alleged sole purpose of propagating the selfish genes.

Within the confines of hindsight of life as we know it, Dawkins can state that 'whenever in nature there is a sufficiently powerful illusion of good design for some purpose, natural selection is the only known mechanism that can account for it'. Indeed, this process of natural selection is deemed by Dawkins as sufficiently powerful to produce organisms whose 'evolvability' is so efficient that 'they save natural selection from wasting its time exploring vast regions of all the possible phenotypes *which are never going to be any good anyway*' (my emphasis).

We are all monsters now

In developing a counter argument to this 'fallacy of perfection', we should not fall into the trap of discussing whether life, as we know it, is of a good natural design, producible by selection. If we start with the assumption that life's functions are 'improbable perfections', then phenotypic self-selection, generation after generation, could produce the desired effect.

A more interesting challenge is to ask what life's functions might have been like if species diversity had taken different routes whenever novelty was produced. There is only one tree of life; it happens to occupy, in the limited time available since the origin of life, only a minute fraction of the multidimensional space of all possible genotypes or phenotypes. I've tried to illustrate this in the accompanying diagram.

It is illegitimate to argue that the supposed harmoniously functioning living 'designs' as we now see them are Dawkins' 'improbable perfections' engineered solely by nonrandom selection. It is also illegitimate to describe the so-called perfections as improbable. We have only one tree of life – it cannot be assigned probabilities. This statistical failure is particularly pertinent with reference to the one-off, passive nature of selection described earlier.

For example, we can describe, with hindsight, how the evolution of the human eye took a known series of steps from the earliest light-sensitive molecules on the surface of living cells to its current form in ourselves. We can also assume, for the sake of simplicity, that at each of the steps there were only two choices: the right choice, as defined by what we now know was the right choice, and the wrong choice, the evolutionary path never explored. Armed with this knowledge, however, it would be a great mistake then to calculate, as Dawkins has done, that the chance of making all the right choices, at each and every step in the known sequence of events, is 1 in 2 to the power of the total number of steps. For example, if there were 1,000 steps, we would have a chance of 1 in $2^{1,000}$ (that is, $2 \times 2 \times 2 \ldots$ for 1,000 multiplications) of taking the right turn for all steps. This wrongfooted procedure leads to the fallacy of the 'improbability' of stringing together the correct sequence of events by chance, which in

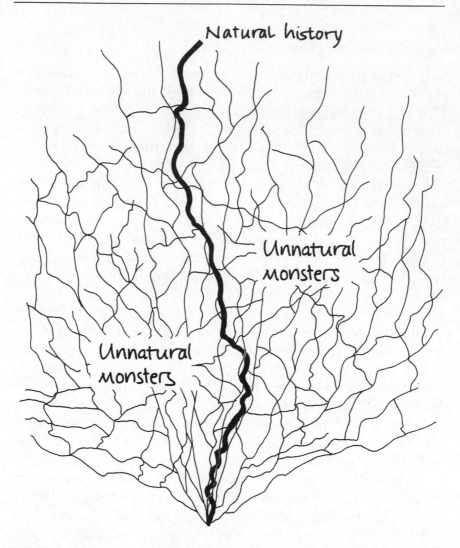

The tree of trees *There is only one tree of life encompassing all known organisms from the origin of life to ourselves. Let's collapse this huge multibranched actual tree of life to one thin line, representing all of natural history. This thin line is then embedded in the 'tree of all trees': that is, the imagined tree that encompasses all shapes, and even behaviours if we wish, of all forms of life that might have occurred if evolution had taken slightly different turns on its journey through the actual tree of life. There is a pervasive assumption that all such imagined other thin lines through the 'tree of all trees' would have contained unnatural monsters not easily explicable as adaptations produced by natural selection. This comes back to the problem of the deep-sea fishes in my earlier diagram. Familiarity breeds contempt for what might have been. There are alternative ways of thinking about the unfilled space of our 'tree of all trees' in the context of both evolution by natural selection and evolution by non-Darwinian methods.*

turn leads to the belief that only selection can bring about the 'improbable'.

This miscalculation of Dawkins, looking vertically through time at the 'known' evolution of the eye, is precisely equivalent to the miscalculation made by Sir Fred Hoyle, the astronomer, when he looked horizontally at the overall sequence of amino acids that makes up a typical protein. Hoyle calculated that, if an average protein consists of, say, 1,000 amino acids and there are twenty alternative types of amino acid to choose from by chance at each position in the protein, then there is only one chance in $20^{1,000}$ to get it right. This is so improbable that Hoyle famously said it would be equivalent to a hurricane throwing together a Boeing 747 from a random collection of scrap iron in a junkyard.

Interestingly, Dawkins recognizes that randomly putting together a particular arrangement of junk in a junkyard is as improbable as putting together a Boeing 747. So to rescue the Boeing 747 as being functionally special, the product of non-random selection, he needs to assume that it is an improbable 'perfection' superior to all other arrangements. But the metaphor is badly chosen when we consider biological structures and functions, and Dawkins' conclusion that all the current wonders of the eye would never have been achieved by 'almost all random scramblings of the parts' is blessed with hindsight.

With this 20–20 hindsight, he is right to state of the eye that 'there is something very special about the particular arrangement that exists'. But he is wrong to assume that all other particular arrangements, which he admits are equally as improbable, would not produce useful devices for seeing – that would be true only if we wanted to achieve the human eye as we know and love it.

Not improbable and not perfect

Even were we to assume that natural selection is the only show in town, we need to ask what would have happened if, say, at some intermediate step in the evolution of the eye, the 'wrong' choice had been made ('wrong' with reference to what we know did happen). Let us suppose that this 'wrongness' resulted in an eye on an insect's leg or antennae. This is not as far-fetched as you might think, Mr

Darwin, because it has been achieved experimentally through manipulation of the genes involved in eye development in the fruit-fly *Drosophila*, as I will explain at another time. With hindsight, we know that this would have scuppered the observed, real history of the eye that ended up with it on the head. With hindsight, we can classify such an eye-on-a-leg phenotype as monstrous and unnatural – ready for the chop by the relentless requirements of selection. However, and this is a very large qualification, we cannot make any predictions from the comfort of life as we see it today of what life might have been like had eyes on legs occurred. There might very well have been some selective advantage to having eyes on other parts of the body from which all subsequent contingent events would have flowed, leading to contemporary life forms with eyes everywhere but on their heads. They would now be described as equally wondrous 'improbable perfections' as eyes on our heads.

The same could be said about the diverse range of functions of the eye: for example, the evolution of an eye capable of using a spectrum of electromagnetic radiation that allowed us to see in the dark rather than perceive 'visible' light. This would not have led to life as we know it, but it would have led to some probable function: not 'improbable' and not needing to be defined as 'perfect'. We need to avoid, when armed with hindsight, making a laughing-stock of all the supposed waste-of-time monsters occupying the unfilled phenotypic space.

What we can say about eyes applies equally well to many of the basic components of our internal cellular and developmental machinery – the genetic code, protein assembly, membrane transport, energetics, muscles, nerves, consciousness, sexuality, language and so on. I once tried to put this point across in my inaugural lecture. I wanted to find a way that was guaranteed to divide the audience on what is natural and what is unnatural. So I began my lecture with a recording of the glorious soprano Maria Callas singing 'Vissi D'Arte' from Puccini's opera *Tosca*.

Let me say at the outset that Maria Callas is the love of my life: her swooping, lyrical, snarling voice brings out the drama of life and the tears from my eyes. I have no resistance to the naturalness of the effect. But for others the operatic voice of Maria Callas is artificial, unnatural and a pain on the senses. Are they beyond rescue? Could familiarity and circumstance sway the disbelievers from the unnatural to the natural? Either way, it is all a question of the perspective of experience. I have much more to say about this later.

It has been possible to sustain the anti-Dawkins argument while remaining within the confines of evolution by means of natural selection. Were I to bring in neutral genetic drift and molecular drive, we would move far away from the intolerant functionalism of the lock-and-key imagery of selection. As I briefly said earlier, biological functions are a complex mix of adaptation, exaptation, molecular coevolution and adoption arising from the reactions of turbulent genomes in turbulent environments.

The forces that underlie our new evolutionary perspectives are tractable but not reducible to the supposed primacy of the freewheeling 'masters of the universe' – the so-called selfish, eternally self-replicating Mendelian genes.

As ever

Westminster Abbey

My dear Dover

Before I begin, could you clarify for me why you sometimes sign yourself Gabriel and sometimes Gabby? I've been going through some of your research publications that you kindly sent to me and I have noticed your use of two different names. Which is your true name? Is there some subtle distinction in usage, or is Gabby to Gabriel as Chas is to Charles?

Well, from your last letter, it's not Chas and Gabby we should be concerned about, but Dick Dawkins. I hope I don't sound as if you have put words into my mouth, but I do find it extraordinary to read your account of his claims about biological evolution. I will, naturally, check out what he writes in his books. From what you say about his popularity, they must be in my local library. I hope that they are not filed under 'Science'! His error over the 'paradox of the individual' is particularly worrying in the extent to which an initial false assumption about the unit of selection can lead to such follies. Maybe I shouldn't use that word, but my irritation is nearly as great as yours. The other disturbing notion is his metaphor of the 'improbable perfection'. Biology is a strange and messy business and 'perfection' seems to be the last word one would use to describe how organisms work, particularly for anything produced by natural selection.

I do, however, have several questions for you. Maybe you have taken his idea of the selfish gene a little too seriously. Maybe it is, after all, just a metaphor for the genes as the units of evolution rather than the units of selection. His mistake is to confound units of evolution (that is, the genetic units of inheritance) with units of selection. Surely, he deserves a little charity and should be forgiven for this slippage. There's no doubt that the 'selfish gene' is a convincing phrase to use if one wants to maximize the extent to which an idea can infect many other minds, whether it is true or not. As we've discussed, I wasn't a bit above that in exploiting the misleading phrase of 'natural selection' to get my point across, and I suspect your 'molecular drive' has something of the propaganda about it. I recall telling Lyell how sorry I had been for my publisher, John Murray, who had doubts about the term 'natural selection' before the book was published. We are all in the same boat, so to speak, at least on this score.

Another question I have concerns your discussion of probabilities and whether there is any logic to life as we know it. Surely life cannot proceed in ignorance of the laws of physics and chemistry. This was certainly true in my day and must be an even more exacting requirement in what I should imagine is a more detailed understanding of physics and chemistry at the molecular level. There is an uneasy sense in which you seem to be suggesting that life could have taken many different routes in the multidimensional space of possibilities: that is, a different turn might have been made at each of your imagined 1,000 steps that led to the evolution of the eye. Surely many of these alternative routes would have petered out for one reason or another, as life came up against the immovable objects of physics and chemistry. By way of contrast, what we can *be sure about is that the 1,000 'right' decisions that were made in the evolution of the eye as we see it before us did not come up against such an impasse, otherwise we wouldn't be discussing the eye. I don't think that you are really suggesting that 'anything goes' on the grand scale of things; nevertheless, I would like some verbal relief from you that this is not your intention.*

I recall writing about this issue, around 1860, to the American biologist Asa Gray, who believed that natural selection could preserve only that which was already designed from On High. I wrote saying that I saw no necessity for the belief that the eye was expressly designed, without however concluding that everything, especially Man, is the result of 'brute force'. I believed, and probably still do, 'that everything, especially Man, is resulted from "designed laws" with the details, whether good or bad, left to the working out of what may be called chance'. Strange how the exact words come back to me.

I confessed to Gray how I managed to avoid feeling cold all over when thinking about the eye; but the sight of a feather in a peacock's tail, whenever I gaze at it, still makes me sick.

I realize that in your more agnostic or atheistic times 'designed laws' might sound arcane and unscientific, but is it just possible that we can substitute the laws of physics and chemistry for 'designed laws'? More importantly, are you any nearer understanding how the 'eyes' on a peacock's tail might have evolved? Is my natural selection sufficient to explain every last detail?

Finally, I still can't get my head around the issues of modularity, redundancy and tolerance that play such principal roles in biological evolution as you see it. I recognize that I'm impatient and that you cannot explain everything in one go, no more than I could satisfactorily give all the evidence for natural selection in one chapter of The Origin. *Furthermore, once you have explained these issues I would rather like to know how they relate to the sorts of things that worried me: for example, the origins of the species, the gradual versus jumpy mode of evolutionary change and the evolution of sex. Are you any nearer solving these 'problems'?*

I think the time has come, Gabriel, to get down to brass tacks, as they used to say in The Potteries, *where I was born.*

I remain your servant

Chas. Darwin

PS. I well remember receiving Karl Marx's heavy tome of a book. I wrote to him immediately thanking him for his 'great work' although I only managed to read the early sections with my German dictionary on hand. I told him that there were similarities between my theory and his 'deep and inspired theory of political economy' – not in detail or mechanism, of course, but in their broad sweeps. I recall saying that 'we both earnestly desire the extension of knowledge, and this in the long run is sure to add to the happiness of mankind'. I did not leave any comments in the margins as was my habit so I must not have taken any deep interest. As with Mendel's work, maybe I should have been a little more attentive. But it's easy to say that with hindsight.

IS DAWKINS AWARE OF THE ERROR OF HIS WAYS?

Dear Charles

I'm beginning to think that it was not some mysterious micro-organism that laid you low and feeble for the rest of your life, but your incessant need to write book after book on what poured out of your learned yet fevered mind. No wonder you had palpitations, bouts of vomiting and severe stomach cramps. No amount of travelling and wallowing in your hot and cold 'hydropathic establishments' (as you called them) could cure you. What you failed to realize, I'm sure, is that the only cure that would have saved you was to stop writing so extensively!

Why am I so bold as to suggest this? It is because since the start of this correspondence, for which I take full responsibility, I have suffered the indignity of the removal of a cyst on the epididymis of my left testicle; I have been subjected to a brain-scan using the new-fangled machine of nuclear magnetic resonance (don't ask) in the hunt for the cause of some mysterious vision problem; and this week my writing hand was completely out of action after a length of dry spaghetti imbedded itself about half way up the inside of my thumb nail! Why do we suffer for our passions? There's no need to answer that.

Genetic bookkeeping

After the publication of *The Selfish Gene*, Dawkins came in for much criticism. It was pointed out to him, time and again, that the description of natural selection in terms of unitary selfish genes confuses the consequences of selection with its root causes. In the days when genes were just mysterious objects floating around the black box of the organism, mathematicians had a field day modelling evolution in terms of abstract units of inheritance, with letters of the alphabet given to each gene. So gene *A* could coexist with gene *a*, where *a* is a mutant form of *A*. Because each human inherits around seventy thousand pairs of genes, however, it is not possible to model the evolution of each and every gene, one at a time. This is particularly problematic when all the genes are interacting with one another in a bewildering array of permutations. One gene can contribute to many different structures and functions, and any given structure and function is built by many different genes. Throw in a great deal of redundancy (many genes existing in multiple copies) and modularity (many genes existing in bits, each of which can be shared by unrelated genes), and we have a considerable problem of representation and modelling on our hands.

Modelling evolution is not unlike modelling the weather or economics or history. The algebraic uses of *A* versus *a*, *B* versus *b*, *C* versus *c* and so on were simply a necessary way around a very difficult problem. Another mathematical simplification is the assumption that genes are additive in their effects on phenotype, not interactive. The use of such additive symbols was therefore simply a bookkeeping exercise. They were never intended to replace individual phenotypes, with their large collections of interactive genes, as targets of selection.

The consequences of this type of bookkeeping were threefold. First, it implied that the gene is the relevant unit of selection, in that models busied themselves with monitoring the rise and fall of *A* versus *a* and so on. Second, it gave the impression that there is a convenient gene (or symbol) for every favourite bit of phenotype that takes a geneticist's fancy. Chop up the phenotype into as many bits as you like (metabolism, proteins, eye lens, brain, consciousness, behaviour, language, intelligence and so on) and out pops a gene. The complexity of individual development became irrelevant, and the intimate cause-and-effect link between development and evolution became a minor irritation that hardly featured at all in the increasingly sophisticated mathematics that clogged up the notebooks. Third, Dawkins turned the symbols themselves into the units of selection and gave rise to an industry of misappropriation of your selection theory, without so much as a blush as to the truth of the matter.

Good cop, bad cop

Does Dawkins believe in his own rhetoric and metaphors? Like all cautious scientists, he has a soft side and a hard side. This ability to face both ways at once was neatly illustrated in his second book, *The Extended Phenotype*, by his use of an image of a Necker cube on the dustjacket. This is a drawing showing the twelve edges of a transparent cube, so that the three-dimensional image can be seen as facing either to the right or to the left. Both left and right images are optical illusions, of course, as the diagram is completely flat on the two-dimensional page. Armed with this ambiguous figure, Dawkins writes that the view of life as evolving through the selection of a selfish gene is just one of the two ways of visualizing the orientation of the cube. The other orientation represents the view of life as evolving through the selection of individual phenotypes. Which way one chooses to view life depends on the problem in hand. The first 'genic' orientation represents his hard line, emphasizing his attempt at scientific originality, whereas the second orientation represents his soft line, reflecting his good old-fashioned belief in Darwinian individual selection all along. So selfish-genery truly has become an optical illusion: now you see it, now you don't.

On this basis, it would seem that Dawkins does recognize the weakness of this theory. Nevertheless, from the worst excesses of the 'paradox of the individual' and the suggested desiderata lists of his 'selfish cooperators' used to solve this pseudo-paradox, it is hard to take comfort from his occasional nod in the direction of orthodox Darwinian selection. Indeed, in many later books, and frequently in the writings of many of his followers, the unitary selfish target of selection has been elevated from a metaphor to reality.

It is therefore not possible to answer your question directly, Mr Darwin. Dawkins writes that he has brought about a 'transfiguration' in the way we view the genetic motor of natural selection through the eyes of selfishly self-replicating genes. He knows that this leads to theoretical cul-de-sacs from which he has to retreat if he is to be taken seriously. But the retreat simply leads him back to the

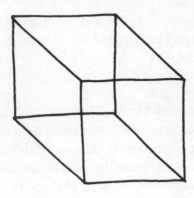

Necker cube. *There are two three-dimensional orientations of this two-dimensional drawing which oscillate from one to the other as you stare at the cube.*

same Darwinian starting point we can all agree on. If there is any originality to be gained from this starting point of individual phenotypes, it is to be laid firmly at your door, and not that of Dawkins. Your theory and your emphasis on the individual do not need a 'transfiguration' into the illogicality of selfish-genery. Dawkins is not alone in assuming that the gene is a self-replicating and ultimately selfish unit of selection. This illusion has often become the starting point for many other theoreticians such as John Maynard Smith, Robert Trivers and William Hamilton, and underpins the philosophical stances of Edward O. Wilson and George C. Williams, among others, on issues of genetic determinism. You might be interested to read what I think about genetic determinism and free will in a later letter.

Selfish-genery is not the same as selfish DNA

There are vast excesses of DNA in the genomes of all species above the bacteria. This is often referred to as 'junk' DNA because it has no clear function, in that the total amount of DNA required for genes and the genetic elements that control gene activity makes up only a tiny fraction of the genome. For example, humans carry enough DNA in each cell nucleus to code for three million genes, if all the DNA were used to produce proteins. In reality, we need only about seventy thousand genes to support our human biological condition.

It was an unfortunate historical accident that some of the early researchers into 'junk' DNA called it 'selfish DNA'. The reason for this is that the excess DNA is made up of thousands of families of identical or nearly identical copies. Clearly, the internal mechanisms of turnover are capable of churning out many copies of the same stretch of DNA, and one could legitimately characterize this in terms of human behaviour by saying that these stretches of DNA are 'selfishly' propagating their own type.

A serious problem concerns the persistent confusion of selfish DNA with selfish genes, and the supposed rescue of the latter by the former. Dawkins' original metaphor was not primarily about the excessive replicating features of junk DNA or about families of repetitive DNA, some of which do actually consist of functional genetic units – for example, the so-called multigenic families and the redundant regulatory units controlling gene activity. Instead, his metaphor applied to all those thousands of Mendelian genes, for each of which we carry only two versions (one from each parent). These genes, as originally considered by Dawkins, were not supposed to be making extra copies of themselves in the lifetime of an individual. They were supposedly capable only of replicating in step with chromosome replication. Their selfishness was an expression of what they did to phenotypes to get themselves excessively propelled into the next generation.

Furthermore, when Dawkins conjured up the selfish-gene metaphor, he was not considering increases in gene frequency by genetic drift or molecular drive. The selfish-gene concept is well past its sell-by date.

Do the laws of physics and chemistry get in the way of biology?

I take your point that there are some regions of multidimensional space that are simply impossible to imagine as capable of being filled by biological organisms that live, breath, eat, reproduce, write letters and so on. Indeed, you are perfectly right to say that such regions might verge on areas where the laws of physics and chemistry might not apply. But the multidimensional space I referred to was a much smaller domain of organization than the one you are alluding to. I am starting off with the assumption that life has already begun, that it is based on carbon (and not on silica, for example) and that it consists of systems of energy transfer and information transfer as we currently observe them in contemporary life forms. Yet, even with such fixed modes of organization, growth and reproduction, it is possible to imagine a multidimensional space of organisms that might have existed at each theoretical branch point in the 'tree of trees', representing a myriad of instances of passive, one-off selection in the long march of evolution.

Let's collapse the actual, multibranched, single tree of life as we currently recognize it into one thin line stretching from the origin of life to humans, and place this line within an imaginary multidimensional 'tree of trees' encompassing all the other collapsed thin lines that might have occurred, but which simply did not occur. I would like you to look again at the diagram I sent in my last letter. In all reaches of this tree of trees, I am still assuming that life is obeying the laws of physics and chemistry. Life cannot escape the forces that shaped, and continue to shape, our Universe. However, within that very broad constriction, there are unimaginable different life forms (including eyes on our kneecaps, or eyes that see in the dark) that might have occurred, and these organisms might have been no worse off for it, or even considerably better off, with regard to their need to grow, develop and reproduce. All three mechanisms of evolution (selection, drift and molecular drive) are opportunistic yet can lead to tolerable and coherent functions.

While on this subject, I want to raise the issue of 'biological laws'. Are there any such things comparable to the laws of physics and chemistry? You speak about 'designed laws' as some halfway house between a deity and what you call 'brute force' (or the 'workings of chance'). This sounds to me as if you are referring to the existence of 'biological laws' that constrain what can and cannot be designed.

There are two ways of thinking about this. If you mean that evolution is nothing but an historical process, with each new stage contingent on the last stage, then I do not believe that there is any argument between us. Life is very much to be seen as a series of 'frozen accidents', as first argued by Francis Crick. For example, once the genetic code has arisen, all subsequent life will need to use it. It would seem quite perverse to invent the wheel all over again. So organisms get locked more and more, through their life's processes, into the myriad instances of problem solving that have occurred over evolutionary time. If this problem solving is solely in the hands of natural selection, then, as I wrote earlier, the view of evolution as ever tighter fits between locks and their keys is apposite.

Slowing down evolution?

Naturally, if the lock-and-key image is correct, evolution will eventually get harder and harder, as contingencies accumulate, despite subtle changes in the environment or in the molecules. As you know, the long evolutionary stasis of biological form in many diverse types of animals and plants is a phenomenon that needs explaining. Maybe one of the root causes of stasis is just this very issue of highly evolved specificity between locks and their keys. Maybe only catastrophic changes in the environment, or in the genetic material, can release organisms from such static structures to explore further the multidimensional space of possibilities.

Recently, the whole issue of evolutionary stasis has become a central feature of a view of evolution termed 'punctuated equilibrium' by Stephen Jay Gould and Niles Eldredge. The idea is that for most of a species' duration, which in some cases is millions of years, there is little evolutionary change apart from the occasional sudden burst. As you will no doubt recognize, such an alternating pattern of stop–go–stop–go evolution rubs up against the gradualism of natural selection that you emphasized so much. I have several things to say about the basis of stasis from the perspective of molecular drive, but I will need to wait until I've completed my account of this process for you in subsequent letters.

So I hope you will understand my position that there are no biological laws as such, but that organisms are subject to the natural constraints that emerge from any historical process. Karl Marx tried to devise a generalized law that governs the evolution of economics and social relations. His method was sound in its broadest of applications to the point that it became generally true and universal, but hopelessly wrong when it came to the fine details of local change at any given moment. Naturally, the evolution of one society from another depends on what has been achieved (economically, culturally, socially and so on) in earlier times, but this dependency does not necessarily imply that there are 'rules' governing society's evolution. Similarly, biology is an historically constrained phenomenon that does not obey any special rules, but which is still capable of incessant diversification and evolution.

Finally, I promise to alleviate your confessed 'sickness' over the sight of a peacock's tail with an explanation of the wonderful world of genetic signals that contribute to eyespots on butterfly wings. But there is much to say before I can reach that stage.

As ever

Westminster Abbey

Dear Gabby

(I would not have dared use first names even to my closest colleagues, other than family, in my day.) I've got some exciting news for you. While I was waiting for your last letter, I inadvertently overheard some agitated mutterings from my close neighbour in the Abbey, Isaac Newton. He was reading aloud from a letter he had received from some physicists who would be contemporaries of yours, the contents of which seemed to have really put his head in a spin. Apparently, all our talk about the constancy of the laws of physics and chemistry from time immemorial is not as solid as we would like to believe. I'm not too sure at this stage what has been said, except that Sir Isaac has been giving me some strange looks over the past few days. He's been particularly concerned to hear about natural selection (after all these years of cohabitation!) and murmuring darkly about the laws of physics having evolved from earlier universes in which they were very different. He is a bit vague on the details at the moment, but he promises to take me through the whole story as soon as he's clarified these strange new ideas with his correspondent. I will, of course, let you know what's afoot as soon as I find out. I hope I can understand it!

At first blush, it does look as if biology has got something to teach physics, and not the other way around. Biology envy instead of physics envy? Who would have thought, if it turns out to be the case, that my natural selection is responsible for everything else in our material world apart from biology? Where does that leave your molecular drive or neutral genetic drift?

Well, I mustn't get too carried away, especially since I haven't heard the details. But from the furious mental scribbling that seems to be going on in Newton's head, I think that something serious is happening. In the interim, let's stick with biology and return to the issue of 'biological laws'.

There is no essential difference, as you say, between biology and history. There is no point in my repeating the points you have made. Nevertheless, there are two ingredients to the history of biology that could be said to be as near to having 'biological laws' as we will ever get. One concerns the issue of 'gradualism' versus what you have referred to as 'punctualism', and the other concerns natural selection itself.

First a word on gradualism. I understand from your earlier remarks about the hopelessness of the 'hopeful monster' syndrome that you well recognize that it is a central requirement of all evolutionary processes to show how the first individual bearing a novelty maintains full biological compatibility with the rest of the population. This requirement is crucial, of course, for sexually reproducing organisms. I have therefore always maintained that only novelties of a very small magnitude (micromonsters?) would ensure that biological compatibility were maintained. As such, evolution by natural selection is gradual and not saltatory: that is, it does not occur by large evolutionary jumps in the form or behaviour of an organism.

If you agree with the above, you will understand my surprise to hear your brief descriptions of 'punctuated equilibrium'. I recognize that you have emphasized the stasis (or equilibrium) part of this phenomenon rather than the 'punctualism', and I look forward to your explanation of the 'basis of stasis'. Nevertheless, if 'punctualism' is in any way denying the essential ingredient of 'gradualism', I don't see how it can be considered a realistic process of evolution.

All of this is a roundabout way of saying that I think that the essential requirement of gradualism in any proposed process for changing the average genetic composition of a population (modification by descent) elevates 'gradualism' to a law of evolution. I see it as so basic and universal a necessity that it governs how evolution takes place. Would you agree that this might be close to being a constant biological principle with accepted levels of predictable outcomes?

Similarly, I want to suggest that natural selection too is close to being a biological law. Could we not accept that, where there are natural differences between individuals that affect their relative reproductive successes and when the genetic contributions to such differences are passed on from one generation to the next, natural selection will always ensue? I accept your emphasis that natural selection is the passive one-off outcome of unique individuals interacting with their unique surroundings; nevertheless, whenever the conditions pertain for such an outcome, natural selection can be said to have occurred. Is this not also a biological law? Not natural selection as a one-off outcome, but the conditions required to bring about the one-off outcome. The preconditions can be said to be universal and law-like with a predictable outcome (natural selection), although the actual outcome in any given instance is not predictable (for all the reasons you've given in your letters).

Well, as you can see Gabriel, I'm doing my best to generate some biological laws. Maybe we should take these issues more seriously now that it seems that physics is beginning to turn to biology for its intellectual sustenance!

There is no need for you to answer these proposals directly – I appreciate that any answers you may have could emerge from the continuing story of modern-day biology and genetics that you are relaying to me.

I am keen to hear, as I said in my previous letter, about what new there is to say about some of our age-old problems of the origin of species and the need for sex, to take just two obvious examples. Our discussion about biological laws can wait until I've sorted out what has got Newton turning in his grave.

Yours affect.

C. O,

GENETIC TURNOVER; OF COURSE, OF COURSE

My Dear Charles

Your last letter is intriguing. I can hardly believe that somebody has decided to write to Newton, as I am to you, and that he is still *compos mentis*. I can't wait to hear from you once you have more details about the evolved nature of the laws of physics. This really does seem to put the cat among the pigeons – although hopefully not among your special breeding pigeons.

I can see that you are doing your best to elevate gradualism and natural selection to a biological law, and you are right to predict that I will have something to say about this. However, as you suggest, we had better not get delayed by these more mundane 'philosophical' issues, but should get back to the more exhilarating matters of what exactly is going on in biology.

Back in 1980 I was mesmerized by Captain Beefheart, the American rock musician and painter. Rock music has some of its origins in Negro slave spirituals in America, which you might have known about. Rumour has it that the young Beefheart began life selling vacuum cleaners to trailer communities in the Mojave Desert. At one trailer, he met Aldous Huxley, author and pioneer user of hallucinatory drugs. Aldous was the brother of Julian Huxley, one of the architects of the melding of Mendelian and Darwinian ideas (the so-called neo-Darwinian synthesis), and both were grandsons of the biologist Thomas Huxley, your own passionate supporter. This unplanned series of connections between Beefheart and you is not unlike the evolution of the genetic connections that make up an organism. Anyway, Beefheart uttered the immortal words, 'this machine sucks' before disappearing to re-emerge as an original rocker.

Beefheart's music was a form of rock that bode no halfway acceptance. You either listened on his terms or you retreated to your previous unenlightened shell. When he commanded you, in his deep authoritative voice, to 'get up and dance', that is what you did, even if it meant pounding the hallowed benches of Cambridge's Sedgwick lecture theatre – named after your own Cambridge master. You confessed to never attending Sedgwick's lectures, which might be the reason he wrote what you called a 'savage and unfair review' of *The Origin* in the *Spectator*!

In 1980 my head was also full of DNA turnover, junk DNA and repetitive genetic elements going round and round in never-ending circles, just like Beefheart's classic: 'Here we go again baby, on the new electric ride; up and down, round and round. With you right by my side; loop the loop, ride and slide, I could barely hold my pride.' The coincidence was fortuitous, but for me it became obsessive. Was there any evolutionary significance to the continual accumulation and turnover of genes, one set of repeats continually replacing another, round and round, side by side? What did it mean that genomes were ten thousand or a hundred thousand times larger than necessary? After all, such gross excess contradicted natural selection's supposed continuous search for gains in efficiency.

Why did so much of the genome and so many of its genes and their internal modules exist in multiple copies? And why were they all subject to such a bizarre variety of non-Mendelian mechanisms, inducing the gain and loss of genetic variants? Surely, there was more meaning to all of this phenomenology than could be had by simply pigeonholing it as 'selfish DNA'. Pigeonholes obscure: we have imagined we have explained something merely by giving it a name, but all we have done is create an excuse not to think about it.

So the song and the DNA went round and round in my head. And then the penny dropped and Beefheart's sonorous ending of 'Bigeyed Beans from Venus' – 'You can be on the right track OF COURSE, OF COURSE' (basso profundo) – took precedence.

What was this penny and why did it clang so loudly? To explain this to you, Charles, I need to spend a little time completing my survey of the turnover mechanisms. In particular, I must explain the chief culprits responsible for the genetic uniqueness of every single human being in the world: the basis of 'DNA fingerprinting'.

If it's your DNA, you won't get away

The story of our ability to analyse the composition of DNA from a single hair or a few saliva cells starts in the Department of Genetics at the University of Leicester. There Alec Jeffreys, now Professor Sir Alec FRS, had the foresight to recognize that some rather peculiar patterns of genetic variation that he had just uncovered could, literally, revolutionize the genetic identification of individuals. Such identification, in turn, would pave the way for the hunt for genes responsible for hundreds of human genetic disorders and, to cap it all, transform the forensic analysis of crime and the experimental genetic analysis of the living world. No individual in any species is immune to being uniquely identified at the genetic level. It is difficult to exaggerate what was at stake with Jeffreys' new technology.

In addition to DNA jumping already described, there are three main genetic turnover mechanisms, often operating simultaneously in the same stretch of DNA. They are called 'unequal crossing-over', 'DNA slippage' and 'gene conversion'. The last one has already cropped up in earlier letters. From a geneticist's perspective, these names reflect precisely what is going on among the DNA sequences, although I well understand that they are a bit of a mouthful and potentially unattractive. But with a little perseverance they can be understood, if my descriptive powers do not desert me completely. It is important to have an overview of how they operate for what I have to say later.

Train swopping

So let's begin with unequal crossing-over. I'm going to start with an analogy. I don't usually like analogies because it is perfectly possible to understand an analogy and still wonder what it has to do with the real thing. Or the analogy can be right and the interpretation of the real thing wrong, and vice versa. But I think I'm on safe grounds in believing that we are all far more familiar, even in your times, with strings of railway carriages that make up a train than with strings of repeated sequences that make up a 'family' of DNA. The connection between the two should be clear very shortly.

Railway carriages can be identified by the company that controls a certain section of the railway system. In your day you had the GWR (Great Western Railway) the LMS (London, Midland and Scottish) and the LMNR (London, Midland and Northern Railway), to name but a few. Today we have Virgin (no one blushes at this name in the late twentieth century), the Robin Hood line and Thames Trains, among others.

Imagine that several trains from several companies happened to enter one central station where the networks intersected, and spent the night there. During the night it was the job of the shunters to prepare the trains for the morning, inspecting the safety of each carriage in turn and removing and adding carriages as the need arose. This was never going to be a very exhilarating job, given the darkness, the antisocial hours and the low pay. Being drunk on the job was an understandable escape from the mechanical dreariness of the work. Drink was all right for the shunters, but not very good for the shunting. When dawn broke there was always the possibility that in a train of GWR carriages there would be one or two LMS carriages. So trains could vary not only in the total number of carriages, but also in the composition of carriages. Making such mistakes was not difficult, even when drunk.

As you can see from my accompanying diagram, two trains (call them A and B) could be lined up side by side and carriages from one used to replace broken carriages from another. If a broken carriage lies near the middle of train A, the most expedient way to replace it would be to decouple the broken carriage and remove it to a siding, along with the other carriages still attached to one end of it. The shunters would then replace all the missing carriages – that is, the broken carriage and its unbroken neighbours – with the same number of carriages from train B. The shunters would be relieved of the necessity of for ever counting carriages in the dark, if they simply aligned the two A and B trains alongside each other, so that both started and finished at the same points. Once a broken carriage and its neighbours on one side are removed, the exact position required to decouple a section of train B in order to replace the missing section of train A can be seen at a glance.

But the intoxicated shunters sometimes make two types of

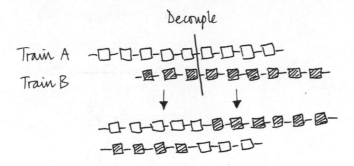

Train shunting (or unequal genetic exchanges) *Train A is lying alongside train B, but not quite end to end. If both trains are decoupled at the same position and then mistakenly joined up again, so that carriages of train B become part of train A, then we end up with two mosaic trains as shown, with one mosaic train inevitably being longer than the other.*

mistake. The first is that they may fail to line up trains A and B precisely. Indeed, only the middle sections of the two trains might overlap with each other, leaving one end of train A protruding on its own on one side, and one end of train B protruding alone on the other side. Now, if, during this unequal alignment, train A is decoupled, say, in the middle and half the train shunted off, and if train B is decoupled in exactly the opposite position as in train A, then train A will receive a section from B that is either shorter or longer than the missing section in train A. The amount of gain or loss of carriages depends on the extent of non-overlap between the two trains, which in turn depends on the state of intoxication of the shunters.

If the shunters have totally lost the plot, they might line up two trains from different companies and in the morning their handiwork would be revealed: mosaic trains with different numbers and types of carriage. Eventually one could, given enough time and booze, replace one company's livery completely with the carriages of another company. When such an event happens between tandem arrays of repeated DNA sequences, we say that one array has been 'homogenized' by another. I'm providing a second diagram to show how this works.

So, all in all, unequal crossing-over has two consequences. First, it continually alters the number of copies of a DNA sequence (or carriages in a train). Second, it can gradually homogenize a set of repetitive sequences with a variant sequence.

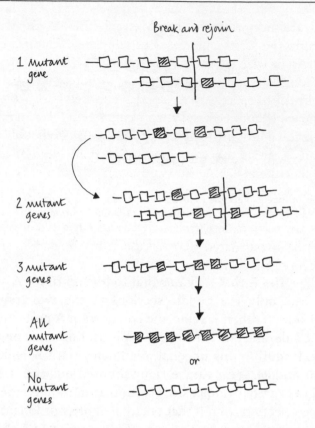

Spreading a mutant gene through an array of identical genes by unequal genetic exchanges
*As with my earlier diagram using trains, two arrays of repeats can lie alongside each
other, but slightly out of phase. A break can occur across the two DNA helices at the same
spot and rejoining can occur as shown. This generates one longer and one shorter array
with the longer array now having two mutant copies of the gene. When this process is
repeated over and over again, we can end up with either an array consisting of nothing but
mutant copies or an array without any mutant copies. An array can become homogenized
for an originally rare mutant gene.*

*What I cannot show in this figure is that during the early stages a rare mutant gene is
more likely to get lost by the continual process of gain and loss induced by unequal
crossing-over. One mutant gene will nevertheless hit the jackpot and ultimately replace
all the original non-mutant genes.*

*One other important feature not depicted in the figure is the spread of a mutant gene
through a population. All the figure can show is the spread down a single chromosome
lineage as the chromosomes replicate and undergo unequal exchanges. If you look back at
my diagram explaining a simple case of molecular drive and substitute unequal crossing-
over between pairs of chromosomes for jumping genes, once again the spreading of genetic
information between chromosomes and the shuffling of chromosomes by sex can lead to
the spread of a mutant gene to all arrays of genes in a population of sexually reproducing
individuals.*

There is another process of turnover called DNA slippage, which has the same two consequences. But slippage differs from unequal crossing-over in that it generally involves repetitive units of only a few bases long (say, from one to ten bases) whereas unequal crossing-over can involve repeats up to several hundred bases long. What's more, slippage occurs between two strands of a DNA double helix, whereas unequal crossing-over occurs between two double helices.

Variation in the number of copies of a repetitive sequence is the basis of the original DNA 'fingerprinting' technique. Our genomes contain thousands of tandem arrays constantly changing in length through unequal crossing-over or slippage. It is technically very simple to measure the lengths of all such arrays in a single individual. It has been estimated that there can be thousands of different lengths of any given array in a population as a whole, so it is highly unlikely that two individuals carry the same lengths.

I'll show you mine, if you show me yours

Gene conversion is the easiest to explain of all the mechanisms capable of homogenizing mutant genes. You may recall that when I was describing the mechanism of jumping of P elements in *Drosophila*, I involved the mechanism of gene conversion at one stage. Remember a P element can jump out of a DNA helix, leaving a gap, and that this gap can be filled by copying the P-element DNA on the opposite chromosome. This is a gene-conversion event. It even works between two DNA helices that lie side by side during meiotic cell division, where one helix carries a mutant gene and the other helix does not. After some jiggery-pokery between the two double helices (which I will not describe), it is possible for the mutant gene to convert the non-mutant gene to its own sequence, or for the non-mutant gene to convert the mutant gene to its own sequence. If we call the mutant form *a* and the non-mutant form *A*, it is possible to end up with either two *a*s or two *A*s when previously there was one of each in an individual.

Let's say we start off with an individual carrying both alleles, *A* and *a*. It would normally produce equal numbers of sperm (or eggs) carrying *A* or *a*. Gene conversion is essentially a mechanism that can create extra numbers of sperm (or eggs) carrying *A* or *a*. So it is clearly a non-Mendelian mechanism. As such, after the appropriate fusion of sperm and eggs, there can be more offspring carrying *A* or *a* in the next generation. Accordingly, gene conversion can contribute to the spread (homogenization) of a variant sequence through a family of sequences and to its spread (molecular drive) through a sexually reproducing population. It is part of the process of molecular drive that alters the long-term composition of DNA sequences both within and between individuals in a sexual population. I explained how this happens at the population level using the jumping (transposition) mechanism as an example.

For the remainder of this letter, I want to concentrate on homogenization and then to return to what is going on at the level of the population.

Genetic homogenization and concerted evolution

In my analogy with trains, we arrived at trains consisting of carriages belonging to different companies, along with the occasional original train of one company replaced totally by carriages of another company. In real DNA, different types of carriage symbolize the existence of mutant repeat units. An array of newly created identical repeats could be expected to decay slowly over time through the independent appearance of mutations in one or another repeat. One of the big surprises in biology during the 1970s and 1980s was the discovery that this did not happen. The repeats remain the same although individual repeats may pick up mutations. What happens is that as soon as a mutation occurs in one repeat it can somehow spread through the adjacent repeats of an array, on a given chromosome, and ultimately to the array on the opposite chromosome. Indeed, arrays of a given repetitive DNA family are often spread over several different chromosomes, and still the mutation can spread to all relevant chromosomes too. You will not gain many Brownie points, Charles, by guessing that the turnover mechanisms have something to do with this.

But the real surprise is that, for a given family of repeats, all arrays on all chromosomes in all individuals of a given species carry the same mutation. At the same time, a different mutant repeat in the same family is shared by all repeats in all arrays in all individuals in

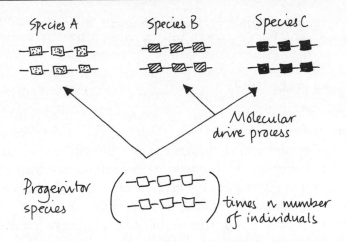

The concerted evolution pattern *Three species, A, B and C, share a family of genes. The genes within a species are more similar to each other than they are to the genes of another species. The pattern of distribution of homogenized, species-specific mutations is known as 'concerted evolution'. Molecular drive is the process primarily responsible for the spread of specific mutations on each branch of the tree, starting from a progenitor species. A number of mechanisms of non-Mendelian turnover power molecular drive. They include gene jumping, unequal crossing-over and gene conversion. I've explained two of these in my earlier diagrams.*

It is not possible to show that each species is represented by thousands of individuals, all of which share a species-specific mutation in a given family of genes.

The repetitive genetic units showing concerted evolution, having undergone molecular drive, can embrace whole genes or be smaller units within the body of a gene. They can also be the small redundant genetic elements involved in the regulation of genes.

another species. As a result, many different repetitive DNA families, including important families of genes, in a given species can be uniquely identified by a specific mutation in the sequences of all the relevant repeats.

This pattern of homogenized species-specific mutations is termed 'concerted evolution' because all the repeats in a given family seem to evolve in unison. I'm attaching a diagram to explain this pattern. How do such homogenized species-specific patterns arise? Before I answer this, I want to emphasise that there are two aspects to the problem. One is to explain the spread of a new variant repeat through a family of repeats in an individual, and the other is to explain how all repeats in all individuals in a population acquire the same new variant. The spread in an individual and the spread through a population are intimately bound up to each other. Together they comprise the process of molecular drive. For the moment, however, I will consider just the spread of repeats within an individual.

Keeping it all in the family

Let's start with homogenization by unequal crossing-over. If you look at my earlier diagram of train swopping, you will see how this can happen. I'm sure you've recognized, Mr Darwin, that if unequal crossing-over continues to produce persistent fluctuations in the number of mutant repeats in each array, there is always the possibility that a newly homogenized array will emerge.

Concerted evolution was first noticed in extensive families of immunoglobulin genes that produce the special antibody proteins that destroy pathogens invading our bodies. Mammalian species contain mutant genes specific to each species that have spread through all member genes of the family in each species. Gerald Edelman, a Nobel laureate for his work on antibody genes, not only noticed this phenomenon but also provided a far-sighted explanation based on 'gene conversion' as one of the underlying homogenizing mechanisms.

As I've explained earlier, gene conversion can take two sequences that may differ by a mutation (for example, *A* allele and *a* allele) and convert one to another, ending up with either two *A*s or two *a*s. There seem to be few boundaries to the operation of gene conversion, in that it can involve members of a gene family nearby on the same chromosome or far away on different, unrelated chromosomes.

Interestingly, gene conversion can be biased in favour of a given gene (for example, in favour of the *A* allele when it is lined up against the *a* allele). It is not hard to see that in such circumstances *A* will spread (be homogenized) through the family of genes at the expense of *a*.

'Give them something to do'

I'm going to end this letter at this point to give you a chance to raise any doubts or queries about the mechanisms of turnover and their involvement in concerted evolution. I admit that there is plenty here to think about.

I recall Gerry Edelman once discussing with me the influence of Noam Chomsky on the world of linguistics. Twenty or more years ago, Chomsky revolutionized our approach to the study of grammar by proposing that language evolved just like any other human feature and that a baby's ultimate ability to make grammatically correct sentences was acquired not through learning and imitation alone, but through the so-called language acquisition device (LAD) active in a developing human brain. No matter what language we acquire, the grammatical rules are considered to be genetically fixed and universal, like the genetic code.

This is far too early to discuss Chomsky's LAD, but I raise the point more in humour, in that Edelman proceeded to tell me that this was a stroke of genius on behalf of Chomsky. Edelman was not entirely comfortable with Chomsky's conceptual proposals; nevertheless, he recognized that Chomsky succeeded to set the agenda for several decades because 'he gave the buggers something to do'. These are Edelman's words not mine, but I understood the point he was making to me at the time. His advice was that, if I were intent on explaining the world of non-Mendelian turnover and its evolutionary significance for developmental and behavioural processes and species diversification, I should start with the description of 'concerted evolution' and then 'let the buggers think about it'.

Concerted evolution is a universal phenomenon and the turnover mechanisms are ubiquitous and common in genomes of bacteria, plants, animals and fungi. Concerted evolution demands an explanation. My proposal is that turnover mechanisms when coupled with sex are directly responsible for concerted evolution. Turnover and sex can satisfactorily explain the dual spread of variant repetitive DNA sequences through a family of repeats in an individual and ultimately through all repeats in a population of individuals. In other words, molecular drive is the process responsible for concerted evolution.

Please don't misunderstand me: I'm not referring to you as one of the 'buggers' of Edelman's brazen approach. I would not be engaged in this correspondence if I did not have the utmost respect for your biological acumen and proven ability to synthesize a unified whole from what might first appear to be disparate items.

I did not take Edelman's advice entirely, in that I have waited to this point in our correspondence before describing the phenomenon of concerted evolution. This is because I wanted to explain some of the background of our 'new biology' first, to bring you up to scratch about genetics. I do indeed intend to spend more time on the links between concerted evolution and molecular drive, but only after I've heard from you.

Oh, and by the way, about my name. I always use the name Gabriel when I write professional scientific papers. By contrast, when I'm more in the mode of expressing an opinion in, say, a review of someone else's research or book, I use the name Gabby. Either name does not stop many French correspondents addressing me as Dear Madame. I never used to like my name Gabby because it reminded me of Gabby Hayes, a long-forgotten and disgusting cowboy with grizzly hair and bad teeth in the torrid films of my youth. Another difficulty was that many long-standing colleagues couldn't get their tongue around it, or around Gabriel for that matter, so Americans consistently call me Gabe, or Gab or Gob.

As ever

Westminster Abbey

Dear Gabby

(*The familiarity comes quite easily.*) *Gerry Edelman certainly has a uniquely expressive style. He could not have got away with such choice words in my day; and you too might have been embarrassed even relaying some of this to me. I hope this correspondence remains reasonably discrete and under wraps! You did warn me, however, in your opening letter how personal and raw scientific discourse can be in your day, with far less modesty and gentlemanly mores. Your anti-Dawkins letter was a bit near the bone, at times, but deep down I must confess that there were some infuriating* biologists in my own day who misrepresented natural selection. I recall writing in 1871 to Thomas Huxley, one of my staunchest and raunchiest defenders, about my exasperations at the persistent attacks on natural selection, particularly from a recent one at the time by St George Mivart in his book The Genesis of Species. I wrote that 'the pendulum was swinging against me and that it would be a long battle for the success of natural selection after we are dead and gone, so great is the power of misrepresentation'. I recall those exact words, although they were written long ago.

Thinking about this now, I am not sure if I entirely agree with Edelman's wisecrack of 'giving the buggers something to do' (I hope my wife, Emma, doesn't hear me using this word; at least she is not lying beside me in the Abbey). It could be said that The Origin was a perfect example of giving them something to do. I marshalled many facts from a diverse range of disciplines and proposed the theory of natural selection. To all fair-minded biologists it was a fairly simple idea to follow, to ponder on and to wonder, as Huxley did, 'why didn't I think of that?' But this wasn't to be. Mivart's powerful misrepresentation was just one of a series of 'objections' that ensured that no constructive analysis and acceptance of natural selection took place in my own lifetime, apart from the chosen few. I admit that they were giants of their time – Hooker, Huxley, Wallace, Spencer and Weissmann, among others – but, in general, the 'buggers' didn't think or do what was expected.

Chomsky's idea about the evolution of universal grammar and LAD sounds very reasonable to me. Maybe in his case he did manage to stimulate a constructive re-evaluation in his own lifetime. There is a danger, however, that some unifying principles can be wrong (as seems to be the case with Dawkins' idea of the selfish gene as the unit of selection). This can lead to a lot of wasteful activity, ultimately distracting scientists from the real job in hand.

Concerted evolution? What would your Captain Beefheart say? 'OF COURSE, OF COURSE'? I have many questions to ask on what I perceive to be the heart of your evolutionary discourse. Primarily, of course, I have to ask what might be the role of natural selection in all of this homogeneity. But before I come to this, some other pressing questions are bothering me.

How do you know that the identical nature of repeated sequences

*is due to homogenization, rather than the weeding out by natural
selection of all mutant repeats that obstruct the proper functioning
of the genes in question? Or that a recent homogenization is not
due to the promotion by selection of a mutant repeat that improves
the fitness of its carriers?*

*Perhaps, indeed, there is a dual process at work. I can grant you
one of your homogenizing mechanisms, such as unequal crossing-
over or gene conversion, as the primary cause of the spread of a
mutant repeat in a particular array on a particular chromosome. But
perhaps we could introduce selection for the promotion of the newly
homogenized array in the population. That is, the homogenized
array, with its genetically identical repeats, could be promoted as if
it were an allele of a Mendelian gene, just as we envisage selection
promoting allele a over allele A, or b over B, etc.*

*You say that some of the repetitive families have thousands of
members that can be spread over several different types of chromosome. How can your non-Mendelian turnover mechanisms cope
with such numbers? Has there been sufficient time to homogenize
all repeats on all chromosomes throughout a species since the time
of its origin? There seems to be such a variety of turnover
mechanisms – do they not get in each other's way?*

The two-step process of evolution

*Finally, to come to the $64,000 question, as you have it, are the
turnover processes simply not just providing additional sources of
genetic variation as grist for the evolutionary mill of selection? As
you are well aware, I always described my process of evolution as
being in two quite distinct parts. The first part has to do with the
question of the generation of fine-grained differences between individuals that need to be heritable; and the second part has to do with
differences in how individuals interact with their environment,
leading to the gradual establishment of an adaptation. I had no idea
of the nature of genes in my day, but this did not prevent me from
proposing a two-part process: mutation proposes, selection disposes.
In other words, the production of genetic variants is a chance event
that is blind to the long-term requirements of adaptations.
Adaptations are shaped by selection not by mutation.*

Is this not still the case with your turnover mechanisms? Do they not just offer a good deal more mutational variation than the simple switching of nucleotides in the DNA? If this is true, they fall within the first and not the second part of my bipartite evolutionary scheme of things.

Concerted evolution is certainly an intriguing problem to think about – it cannot be ignored if your turnover mechanisms are all over the DNA. Nevertheless, are you not getting just a little bit confused about the crucial distinction between the two steps of the evolutionary process? I'm sure you've been asked these questions many times before, but it is important for me to get it straight in my own mind. Unless I do this, we could end up talking at cross-purposes when we move on to more involved issues.

Your most sincere friend

Charles Darwin

MOLECULAR DRIVE FOR ADVANCED PLAYERS

Dear Charles

I have indeed heard all your questions before from my own con-temporaries, but your phrasing of them makes them no less original for that.

I'll take your questions as they occurred to you. Often the fluid, spontaneous order of thoughts reveals just how our minds cope with new information in making essential links with already existing information in our heads. But enough of this self-consciousness; let me get down to the scientific questions.

Your first question considers the possibility that repetitive sequences are identical because selection has been busy conserving particular sequences and not allowing new variant repeats to spread. To answer this and other questions, I'm going to focus on one real example. These are the genes that code for the so-called ribosomal RNA, which is the RNA required for the translation of messenger RNA into proteins. The genes (called rDNA for short) can exist in hundreds, sometimes thousands, of copies in each individual. In the fruitfly there are about 500 copies, 250 on each of the X and Y sex chromosomes. In humans there are about 700 copies, spread over several pairs of different types of chromosome. Usually, but not always, they exist in tandem arrays.

The rDNA genes are one of the oldest genetic families that we know of. They exist in all forms of life from bacteria onwards, and are absolutely crucial for the proper transfer of information from genes to proteins. As such, they are not disposable, and yet in many groups of animal and plant species they are continually homogenized for new mutations. For example, all 500 rDNA copies in the fly *D. melanogaster* share a single unique mutation. Similarly, seven other species that are closely related to *melanogaster* contain mutations that are unique to each species, and yet which have spread to all rDNA genes in all individuals. The same concerted pattern is found in the rDNA gene family of the great apes, including our human species.

Accordingly, we can envisage a situation in which the rDNA in some progenitor species was of a uniform composition. Over time, as new species arise from the progenitor, new variant repeats spread through the family of genes in each individual and ultimately through all individuals of a sexually reproducing species. Despite the importance of rDNA, selection has not prevented the spread of new mutant repeats. Crucially, some homogenized mutations occur in functionally important sequences, yet have not been eliminated by selection. How is concerted evolution achieved throughout a species and what might be the role of selection?

Genetic buffers

Imagine that one rDNA unit on one chromosome in one individual picks up one mutation. In the fullness of time, we know that this mutant rDNA unit will spread and replace all the original non-mutant rDNA units, throughout the species.

It is difficult to imagine that the first lone mutant repeat would affect the viability and reproduction of its individual carrier. Genes that exist in multiple copies are buffered from the effects of a mutation when it first arises.

However, just for fun, let's bring in selection at the point of initiation of the first mutant repeat. Let's assume that the mutation does indeed have a beneficial effect on reproductive success, no matter how small, notwithstanding the buffering provided by multigene families. Eventually, we can expect the chromosome carrying the single mutant repeat to have spread by natural selection to all individuals.

But have we solved our original problem? Can we expect natural selection to encourage the spread of all mutant repeats, one at a time, for each of the 250 copies of the rDNA on the X chromosome and of the 250 copies of rDNA on the Y chromosome in the fly? Furthermore, on what grounds can we expect the same mutation to occur in each of the 500 rDNA repeats, just waiting to be promoted by selection? As you may agree, Charles, this becomes an impossible argument to sustain, even if we were willing to engage in the worst of just-so storytelling.

It is now generally recognized that we do not need to look beyond unequal crossing-over and gene conversion as the two mechanisms of DNA turnover responsible for concerted evolution. There is good evidence for the operation of these mechanisms in the rDNA and most other redundant genetic systems.

An argument for natural selection is even harder to sustain when we are dealing with patterns of concerted evolution in extremely large families of junk DNA that have no apparent function, let alone effect on reproductive success. For example, in the mouse there is a family of sequences that consists of roughly three million copies of a repeat some 240 bases long. Another family, called the alpha-satellite family in humans, consists of several hundred thousand copies spread in tandem arrays all over our twenty-three pairs of chromosomes. Both families show aspects of concerted evolution.

Bringing in selection at a later stage

Maybe selection needs to wait until an appropriate number of mutant repeats have accumulated by molecular drive before an effect on reproductive success takes hold. Biology being what it is, it is simply not possible to say what the threshold number in any given family has to be before natural selection has its way. This can be discovered only through the detailed examination of each multigene family in turn. These families come in all shapes and sizes and perform a multitude of cellular functions, affecting almost all features of living organisms. A high proportion of genes are internally repetitious and the important genetic elements that regulate the activities of genes are invariably repetitious. All these features, including the development of eyes, ears, giraffes' necks and so on, will have evolved through molecular drive and natural selection acting on redundant genetic elements. Our problem is to understand how this takes place.

To consider how selection can get involved with continuous rounds of homogenization, it is important to understand the molecularly driven process at the population level. As you rightly pointed out in your letter, all evolutionary theories need to show how a population of individuals change in their genetic composition: the critical second step in your two-step procedure. Molecular drive cannot escape this responsibility.

Is the chromosome a natural barrier to homogenization?

To understand the population aspect of molecular drive, I need to come back to the chromosome. I accept your suggestion that if unequal crossing-over or gene conversion were restricted to one chromosome, then it would be possible for each array of genes to be separately homogenized. Each chromosomal array could therefore be the target of selection. But we have plenty of evidence that the chromosome is not a natural barrier for the exchange of repeats. During meiosis, when similar chromosomes lie side by side, there is plenty of opportunity for unequal crossing-over. Furthermore, gene conversion was discovered as an event between two alleles of a gene on two similar chromosomes. We now know that all turnover mechanisms can work at several different levels: within a single chromosome, between a pair of chromosomes and between different types of chromosome.

How do we know all of this? I will give you a few examples. Examination of mutant repeats in the alpha-satellite DNA of humans and other great apes reveals a wide range of chromosomal distributions. Some mutations have spread only partially down a single chromosome; some have spread fully down a single chromosome; some have spread to one particular pair of chromosomes; and some have spread over two or three particular pairs of chromosomes; some, indeed, have spread to all types of chromosome of a species, revealing the full pattern of concerted evolution.

Like ripples in a pool

Many mutations spread to different types of chromosome even before they have completely homogenized the array of repeats on the chromosome of origin. It is as if the 'transition stages' of molecular drive are laid before our eyes, like ripples in a pool. The least homogenized repeat variants would be the inner ripple, yet they would be equally represented on all chromosomes, with the most homogenized repeat variants representing the outer ripple.

Evolution by molecular drive is not, therefore, an instantaneous process. It takes time to spread an initially rare variant through all the relevant repeats in a species. The exact amount of time depends critically on the activity of the turnover mechanisms, the facility to spread to other chromosomes and the size of the family of genes. No mouse woke up one day with three million copies of junk DNA clogging up its chromosomes.

Molecular drive at the population level

So far, I have been describing this spread as if it takes place in one idealized individual living in a vacuum, away from all other individuals in a population. This was often the way in which the process was modelled on the computer in the early days. But molecular drive can bring about a 'modification by descent' of a whole population, with the passing of generations – operationally distinct from selection or drift, but with the same general consequences. It was this realization – moving from a single isolated chromosome to many different chromosomes in a sexual population – that brought the molecular-drive theory of evolution on a collision course with my Captain Beefheart lyrics. Round and round in ever-widening circles.

The spread of mutant repeats at the population level is particularly interesting and can greatly influence subsequent interaction with selection. I have called this the 'cohesive mode' of population change under molecular drive. I will conclude this letter with a description of the dynamics of this change and then wait to hear from you again.

All in the same boat

What was exercising my mind back in the 1980s was the involvement of natural selection in homogenization. To understand how these two procedures interact, we need to come back to the chromosomes.

Without turnover between repeats located on different chromosomes there is no molecular drive. But not only does turnover frequently occur between the chromosomes, the sexual process ensures that they are continually being shuffled. Hence, through repeated cycles of turnover between chromosomes, and the sexual process ensuring that chromosomes separate and enter different individuals, we arrive at the molecular-drive process, capable of inducing long-term changes in the genetic composition of a population.

An important feature of molecular drive is that the rate of sexual mixing of chromosomes between individuals is much faster than the rate of turnover between chromosomes. An important consequence of this is that the extent of homogenization among individuals, living in any given generation, will be about the same. All individuals are in the same boat. If we are expecting about 30 per cent of a family of repeats to be homogenized in a given generation, then each individual living in that generation will be homogenized at about the 30 per cent level. No one individual is left behind with no variant repeats at all, and no individual races ahead with nothing but fully homogenized variant repeats.

Under molecular drive the overall genetic uniformity, or cohesion, of a population is maintained throughout a period of change. Naturally, there will be some distribution of individuals with more or less homogenized repeats around the average 30 per cent mark, or the average 50 per cent mark, and so on. Nevertheless, the distribution of individual genotypes around a given average will not extend from one extreme of zero variant repeats to the other extreme of all variant repeats.

This feature of a cohesive population change under molecular drive governs any future interaction with selection. When a molecularly driven change in repeats gradually embraces all individuals of a sexual population, selection can do two things: either stop the process continuing altogether or ensure that there is some internal adjustment that accommodates the change. The latter possibility is called 'molecular coevolution', a process that I introduced earlier when I described the selection of genetic systems capable of repressing the terrible effects of a molecularly driven accumulation of P elements in fruitflies.

In my next letter I will tell you how molecular coevolution works in the rDNA gene family. Then I will take two particular examples of multigene systems that are making the headlines at the moment. One concerns the so-called multiple *Hox* genes that control many features of development in animals, and the other concerns the so-called *period* genes that control biological rhythms and courtship behaviour in animals. I also promise not to forget the butterfly eyespots.

But before I enter into this, I would like to hear back from you. I fully accept your two-step process of evolution. But I hope I've explained how molecular drive represents the second step of aiding the spread of mutational variants and is not just involved in the first step of producing these variants.

With best wishes

Gabriel

Westminster Abbey

My dear Dover

I've enjoyed this last letter of yours, although I suspect that you might inadvertently spawn a hundred different versions of molecular drive, as I did with natural selection. It seems to me that you have two uncontroversial facts in your favour: the phenomenon of concerted evolution and the existence of all your non-Mendelian turnover mechanisms. The crucial issue is not, therefore, the existence or not of homogenization of families of repeats by variant repeats, coupled to the ultimate spread of the variant repeat through a sexually reproducing population. Rather, it is the role that natural selection might play in encouraging or discouraging molecular drive in any given gene family. I suspect that you will deal with this issue when you describe 'molecular coevolution'.

I can accept now that molecular drive cannot be reduced to just a new source of mutational grist for the selection mill. If there is to be any involvement with selection, then perhaps selection needs to wait until a given degree of spread of a variant repeat has taken place. As you say, a redundant system of genes is initially buffered against the presence of one mutant member. Moreover, the interesting way in which the genetic cohesion of a population is maintained, throughout a period of change, would greatly influence how we think about selection in such circumstances.

But before you start with such considerations, I'm intrigued to learn where the name 'molecular drive' came from. The 'drive' part suggests a certain inevitability in the spread of a given variant repeat. For example, the whole point of natural selection is that it is about inevitability. In other words, selection is a non-random directional force in evolution that sorts out the variants produced haphazardly by random mutation. I don't have to tell you that I'm not suggesting that selection knows where it's going, as if it were some force working to some predestined eye, heart or whatever. Nevertheless, there are very good biological reasons why one mutant allele is taken through to the next generation better than an alternative allele.

Dover, I hope you don't mind my being blunt, but I do not see any cause for directionality at all in what you have told me about the workings of unequal crossing-over, slippage and gene conversion. There does not seem to be any good reason why one variant repeat is homogenized and molecularly driven through a population and not another. There seems to be a large element of chance as to which of the randomly produced variants is randomly spread by your process. The 'drive' part of 'molecular drive' might not be a very appropriate word and could be misleading, if you are not careful.

When it comes to mobile elements, such as P elements in Drosophila, *I'm willing to accept an element of drive, in that it is the sequence and structure of a given piece of DNA that influence its mobility and accumulation. I realize that we cannot simply attach the epithet 'selfish' to Ps because they are not autonomous, self-replicating entities, given the extent to which they rely on host enzymes to do anything at all. But setting this aside, jumping elements can be said to be 'driven' in a way that does not apply to your other turnover mechanisms, as I understand them.*

If I'm right about this, maybe your choice of name for the internally activated process of evolution was simply a device to get the process lodged firmly into people's minds. I'm not entirely innocent myself of this way of proceeding – new ideas, if right, need all the help they can get. As we've discussed earlier, 'natural selection' was a clear misnomer and caused a lot of confusion in my day; and from what I hear from you, it has continued to do so for a hundred and fifty years. 'Artificial selection' is so well established, with a clear, obvious and active 'selector' at work, that the use of the word 'selection' as a name to describe the outcomes of unique environmental interactions with individuals was bound to cause trouble. My only defence is that everyone has moved on and understood what I really meant, if we ignore those people who think in terms of genes as the targets or units of selection. Indeed, the 'selfish gene' is a very good, thought-provoking and memorable phrase that presumably has done no harm for the sale of books. As I have cheekily suggested before, Gabriel, are you not engaged in a bit of black propaganda yourself?

Yours in anticipation

MOLECULAR COEVOLUTION

Dear Charles

What's in a name?

A name can be both seductive and provocative at the same time. The seductive side is that it identifies and encapsulates a particular phenomenon that can aid its spread among a population of thinkers. The provocative side is that it forces other people to consider a given phenomenon in the terms in which you propose to identify it. I believe that evolutionary biologists don't usually like new terms, especially if there is an element of admission by the user that there is a real, live phenomenon out there that is worth discussing. This last point has more to do with human psychology and the sociology of science than with science itself. Most geneticists do not easily use the metaphor of the 'selfish gene' because it is seriously at odds with how genes are known to interact with each other in their contribution to the real targets of selection, individual phenotypes. On the other hand, as a result of the seductive advocacy of the book *The Selfish Gene*, and the use of the term among science writers, journalists, general biologists and professionals from unrelated disciplines such as philosophy, sociology and psychology, the term has entered common language. I use it extensively because I need to show just how much at odds the idea of the selfish gene is with our new knowledge of genetics and development. So in a sense I'm also complicit in the continual dissemination of the name.

I wonder what the initial reaction was to the name 'natural selection', rather than to the idea of natural selection. Did it naturally come tripping off the tongues of all and sundry? Or was there some resentment at being forced to recognize your theory through the very usage of the name? We have the same problem with the word 'god'. How can one discuss 'god' without recognizing the name?

Molecular drive, as a name, suffers from all these problems, with an added complication arising from the splintering of biology into mutually incomprehensive subdisciplines. As you might have already realized, an appreciation of molecular drive demands an understanding of turnover among genetic elements. You will see shortly that it also demands a working appreciation of how genes contribute to multicellular development. This is not necessarily an easy subject to grasp if one is, say, a student of animal behaviour. Similarly, molecular biologists are not overfamiliar with the subtleties of ecology and the importance of thinking about evolution at the level of populations and generations, although they are familiar with issues of turnover, redundancy and modularity in genomes and development.

Be that as it may, I did feel that a proper name was required for the process underlying the pattern of concerted evolution. Concerted evolution, in itself, was a good name, but it was in danger of becoming a little like your term 'adaptation', a simultaneous description of both the cause and the effect.

So why molecular drive? 'Molecular' was chosen because I wanted to emphasize that the evolutionary process under review was an outcome of internal, molecular turnover in DNA sequences. The term 'drive' has a different origin and, with hindsight, I would perhaps not have chosen it today. It has led to confusion both with the issue of directionality that you raised in your last letter and with another process with historical precedence called 'meiotic drive'. Meiotic drive is also a non-Mendelian system, but it operates at a different level from the level of the molecular activities that govern molecular drive.

Getting there, without being pushed

First, I want to come to your important point about directionality. As you rightly say, natural selection is a deterministic, non-random product of survival pressures arising in the environment. Mutation is random, but the 'selection' of some phenotypes, with their genes, over others is not random: it depends on how they interact with their environment. You are also right in objecting to the use of the word 'drive', given that the outcome of an unbiased gene-conversion event, or an unequal crossing-over event, is purely random. From an individual with two copies of a gene, A and a, we can convert gene A to a, or a to A, at roughly equal frequencies.

Although it might not seem obvious, even unbiased gene conversion, like unbiased unequal crossing-over, can lead to the ultimate spread of one variant repeat over another. Let's forget multigene families and think of just one gene with its two alternative alleles, a and A. A non-Mendelian gene conversion could go either way, Aa to aa or Aa to AA. It is unlikely, given the inexact nature of biological processes, for there to be complete parity in the direction of conversion so that, after repeated rounds of gene conversion, the number of AA individuals equals the number of aa individuals. Through random fluctuations and accidents of sampling, there could be a higher frequency of conversion events producing AA than producing aa. As such, AA would increase in the population at the expense of aa. Even if in the next generation aa is more frequent than AA and we get a reversal of what happened in the first generation, it is still possible, after tens of thousands of generations, that either A or a will come to dominate the population. This sort of accidental drift due to continual fluctuations in the direction of gene conversion is similar to accidental fluctuations in numbers of sperm and eggs and in sizes of populations that give rise to the evolutionary process of neutral genetic drift.

For the moment, unbiased gene conversion, or indeed any unbiased mechanism of turnover, can lead to the accidental homogenization and spread of a variant gene, through accidents of sampling. It could take a very long time, but it is not impossible. Under such circumstances, the word 'drive', implying a deterministic process, is not a happy choice of word.

Biased systems: we have ways of making you go

There are many times when gene conversion is biased, in that a gene conversion between A and a persistently produces more AA than aa. Under such circumstances, A will spread more rapidly to replace a. In these cases the 'drive' in molecular drive is appropriate. Our problem, however, is that we do not know, simply by looking at the distribution of mutations in a family of genes, how much is due to unbiased gene conversion and how much is due to biased gene conversion. Fluctuations between biased and unbiased gene conversion can depend on the genetic make-up of individuals.

I originally thought of using the term 'molecular drift' for unbiased turnover and 'molecular drive' for biased turnover. But this did not solve my problem. I did not want to suggest terms that might imply that we actually knew the mode of any given turnover mechanism in any given gene system. What's more, there are many examples of redundant genetic systems in which several turnover mechanisms are operating simultaneously among the repeats, and at different rates, with different biases, and on different unit lengths of repeat. For example, many of the arrays of sequences in the human genome, discovered by Alec Jeffreys, provide a unique genetic fingerprint for each individual. They are subjected not only to unequal crossing-over and slippage, as I said earlier, but also to biased and local gene conversion between repeats on different chromosomes. In other redundant genetic systems, such as the hundreds of genes involved in the formation of eggshells in silk-moths, the production of a small array of short repeats by DNA slippage is known to initiate the beginnings of a gene-conversion event.

In view of all of this, and with some reluctance, I went for the generic word 'drive' as in molecular drive. This was primarily to indicate that there are internal genomic processes that can drive evolutionary changes, on a par with the more familiar external processes that promote evolutionary change – natural selection and neutral genetic drift. I see molecular drive as a third mode of evolutionary change to add to the other two. This is the important point to make. Whether the name itself is used is less important than the recognition that the evolution of our genes and their controlling elements is influenced by a variety of turnover mechanisms, and that this has implications for how we interpret the origin and establishment of novelties in form and behaviour. Molecular biologists have no problems with turnover mechanisms, but they tend to leave evolution to natural selection alone. This, of course, suits the evolutionary biologists who are not too willing to leave the comfort of natural selection. There is a long way to go, therefore, before either group freely, and with knowledge, uses the name 'molecular drive' whenever the situation demands.

Molecular drive and meiotic drive are not the same thing

I was well aware of the phenomenon of meiotic drive when I coined the phrase 'molecular drive' because I had been working for several years on meiotic drive affecting the segregation of special plant chromosomes called 'B' chromosomes. In some species there are meiotic processes that ensure that one out of a pair of chromosomes enters more gametes (sperm and eggs) than would be expected from Mendelian segregation. This is particularly interesting because the chromosome is supposed to be the only true Mendelian unit, as I explained in an earlier letter. The non-Mendelian segregation of chromosomes is called meiotic drive, which is a good and appropriate name. Meiotic drive is truly a deterministic process. It is, however, very rare. Most chromosomes in most species segregate into sperm and eggs along strict Mendelian lines. For this reason, I did not want the much more prolific, ubiquitous and variable range of non-Mendelian turnover mechanisms to be confounded with meiotic drive. Meiotic drive is just the tip of the iceberg of what is happening with the bulk of the DNA.

Molecular drive and selfish DNA are not the same thing

In the same vein, I did not want the variety of turnover mechanisms and the constant switching from biased to unbiased modes of operation to be confused with the concept of selfish DNA – which needs to rely, if it is to be believed, on a deterministic process such as transposition or biased gene conversion. I've discussed this before, Mr Darwin. If I'm forced to call DNA anything, I prefer to call it 'ignorant DNA' in the sense that it is ignorant of all the mechanisms of turnover that play on it. Apart from some transposable elements and some instances of biased gene conversion, most sections of DNA in most genomes can be multiplied no matter what sequence they contain. Similarly, our drunken shunters in the railyard can increase or decrease the number of carriages in a train, irrespective of the company livery that adorns the sides of the carriages.

However, I am prepared to admit to you, Mr Darwin, that molecular drive is not a wholly satisfactory name. But it has been out in the literature since 1982 and I cannot think of a better term that would justify changing it. It does not overly concern me whether the actual name is used as such, so long as the complete A to Z of the evolutionary process in question, both internally within genomes and externally in populations, is fully understood. Only then can the interaction with selection be realized in a realistic manner.

But let's get back to some more biology.

Welcome selection

'Houston, we have a problem.' These were the immortal words of the astronauts stranded out in space on that fateful journey in 1970 to the Moon in Apollo 13. Yes, Mr Darwin, we can travel to the Moon and back, although when I say 'we', I mean a small handful of highly trained men and women who contain enough of 'the right stuff' to allow themselves to be hurled towards the Moon inside rockets with the thrust of millions of horses.

In the world of evolution we also have a problem or, more accurately, an intriguing observation that demands an explanation. This is the phenomenon of molecular coevolution, first seen in the same rDNA multigene family that I have been using as an illustration of concerted evolution.

Some of the first questions raised about concerted evolution, and its underlying motor of molecular drive, were along the lines: Should we care about it? Do continual rounds of homogenization have anything to do with the evolution of biological functions and adaptations? Or, if a mutation is inadvertently homogenized and as a result affects the critical part of a gene, then why can't selection get rid of the problem? After all, it gets rid of almost anything else detrimental to organisms shaped solely by the Darwinian adaptive process.

To tackle such legitimate concerns, let's look at the likely range of effects of a newly homogenized mutation in, say, the rDNA family. The easiest case is a mutation in regions that are, to all intents and purposes, inessential sequences. No matter what the degree of homogenization, selection would have no interest in it.

A second case is a mutation in a critical region of the rDNA unit that ultimately improves the reproductive success of individuals. Given the buffering consequences of redundancy, however, we would need to know what proportion of all rDNA units need to be mutant to generate an individual with improved reproductive success. The threshold at which this might happen depends on both the gene family and the species. When the threshold is reached, the cohesiveness of molecular drive will ensure that all individuals reach the threshold more or less simultaneously.

We can suppose that individuals with above average numbers of beneficial variant repeats survive and reproduce better than others. Accordingly, there will be an even higher average number of variant repeats in a later generation than expected if molecular drive were acting alone. With this scenario, we can now expect the dual action of molecular drive and natural selection to promote a beneficial mutation.

The case of the duff gene

So far, I've dealt with the cases of the neutral mutation and the beneficial mutation. The detrimental mutation turns out to be the most interesting of all. If a detrimental mutant repeat is molecularly driven through a gene family and also through a population then we can expect some threshold level at which the number of mutant repeats is too great a burden to bear. Unfortunate individuals at this threshold, inheriting a greater than average number of repeats, would be weeded out by selection. It would be as if the slowly changing population finally came up against the brick wall of selection. Individuals with above average numbers of detrimental repeats are the first to suffer. Ultimately, the incessant process of molecular drive, particularly in its biased mode, would push all the population into the jaws of selection. This would lead to extinction.

The evolution of tolerance

And now comes the surprise. There is a fourth option, centred on the phenomenon of molecular coevolution, which I believe is an important feature in the establishment and maintenance of biological novelties influenced by redundant and modular genetic systems undergoing genomic turnover.

Charles, I'm going to introduce a small acronym for all such systems. It is tiresome for me to write 'genetic systems showing turnover, redundancy and modularity' over and over again, as it must be to read the same. So I'm going to call them 'turnover, redundancy and modularity' (TRAM) systems. When I've finished describing many of the genes involved in development and behaviour, particularly the genetic circuits that control gene activity, you will understand that TRAM systems are more the rule than the exception. I beg your patience while I get back to molecular coevolution.

Molecular coevolution was first observed in rDNA, and its evolutionary significance was thrashed out by me and my close colleagues, Richard Flavell and Enrico Coen, in Cambridge.

In several genera of animals and plants, mutations have been homogenized in functionally important regions of rDNA, for example in the regulatory regions or in the genes themselves. Why were such mutations not thrown out by selection? The answer is that these mutations could be tolerated as long as selection promoted the spread of compensatory mutations in other genes that interact with the TRAM system. Vital biological functions could therefore be maintained despite continual rounds of homogenization and continual fluctuation in the number of redundant elements because the functionally important interacting molecules coevolve.

I will focus only on molecular coevolution in the regulatory regions of rDNA because the evolution of biological novelties has more to do with restructuring the regulatory regions of most genes than with changes in the protein-coding sections of genes.

Regulating genes by TRAM

We call regulatory regions 'promoters'. These are sequences of DNA usually lying next to the start of a gene that can promote the activity of the gene (that is, its expression or transcription). They can also demote the expression of a gene, so 'promoter' is a slightly one-sided term. To help you visualize a typical promoter, I'm enclosing a diagram.

For a gene's promoter to work, it needs to be bound by specific

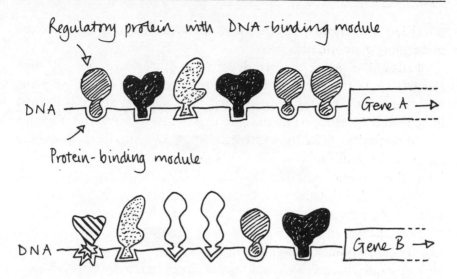

Modular regulatory regions (promoters) of two genes, A and B *Genes are turned on or off by a variety of regulatory proteins that recognize specific binding sites (short stretches of nucleotides) in the DNA just in front of the gene proper. The region of DNA containing a multitude of protein-binding sites is called the promoter. In this diagram I've drawn three different regulatory proteins (shown by different shapes) slotting into the corresponding shapes of the promoter binding sites of gene A. For gene B I've drawn four regulatory proteins and four binding sites. Note two things: first, there can be more than one copy of a given binding site in a promoter; and second, two unrelated genes can share a binding site and its corresponding regulatory protein.*

Such fluctuations in the number of copies and sharing between genes indicate that the binding sites are modular and can autonomously get around the promoters of genes during evolution by one or other mechanism of turnover.

proteins called regulatory proteins. These are, of course, coded by other genes. Each regulatory protein folds into a particular shape and a modular part of the overall shape will recognize a specific module of DNA sequence in the promoter. These DNA modules are called binding sites. As you can see from my diagram, several different types of binding site can be recognized by the corresponding modules of differently shaped regulatory proteins. And any given binding site can vary in number. I've drawn two fictitious promoters of genes *A* and *B* to show that some binding sites are shared. Similarly, some regulatory proteins have several different types and numbers of modules capable of recognizing different binding sites. Although I haven't shown it in my diagram, the DNA-binding modules can, like the binding sites, be shared by different proteins and their genes. I've

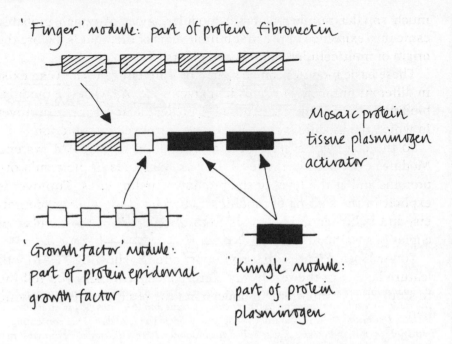

'Finger' module: part of protein fibronectin

Mosaic protein
tissue plasminogen
activator

'Growth factor' module:
part of protein epidermal
growth factor

'Kringle' module:
part of protein
plasminogen

Modular proteins *Proteins are constructed by modular components that can vary in number and also be shared by unrelated proteins. As with the modular promoters in my previous diagram, these two features indicate that modules can independently spread from one gene to another and hence from one protein to another, during evolution, by one mechanism or another of turnover. This diagram shows part of the protein called the tissue plasminogen activator, consisting of three different modules. One module is the 'finger' module of the protein fibronectin; another is the 'growth-factor' module of the epidermal growth factor protein; and the third is the 'Kringle' module of the plasminogen protein.*

 The modular, mosaic nature of proteins exists at the level of the genes that code for such proteins. Regulatory proteins are often mosaic structures capable of binding to a number of different binding sites in gene promoters.

included a further diagram to show the modular nature of some proteins. I'm looking forward to telling you about the 'master' gene (*Pax-6*) that controls eye development in most animals. This too is a mosaically constructed gene with different combinations of DNA-binding modules.

 Perhaps all genes, like their promoters, are mosaics of smaller modules that can spread around the genome. Indeed, there may be as few as 1,500–2,000 separate functional modules. The 70,000–100,000 different types of gene therefore represent different permutations of a much smaller number of basic modules, most of which probably came into existence in the first billion years of life: that is, before the origin of multicellular animals and plants.

These basic modules can be shared by different genes and can exist in different numbers of copies in a given gene. As with the modular binding sites of promoters, these features indicate frequent turnover in which the module is the basic unit of turnover and function.

So the complete system of gene regulation is a TRAM system. Modules exist both at the level of individual folds in the regulatory proteins and at the level of the promoter binding sites. Turnover is explicit in the sharing of modules by unrelated proteins or promoters; and redundancy is obvious from fluctuation in the number of copies of specific modules.

You will see, Charles, that regulatory circuits showing TRAM will feature largely in the rest of my correspondence (unless you tell me to stop), so it's important to understand the basic outline I'm giving here.

The evolution of slot-machines

So what happens if a mutation in a protein-binding module of a promoter causes the DNA-binding modules of regulatory proteins to bind less efficiently?

As I explained above, redundant systems are naturally buffered against these mutations until the threshold of mutant repeats is crossed, at which point the reproductive success of individuals is affected. Instead of putting up a brick wall, selection can now promote the spread of mutations in the requisite module of the regulatory protein that restore efficient binding between the mutant protein module and the mutant binding modules of the promoter. In other words, selection and molecular drive have mutually cooperated in molecular coevolution between two different genetic systems. I've drawn another diagram illustrating molecular coevolution that I hope will help you. You can see that in species X the modules of regulatory proteins match the shapes of their binding sites. The same is true of species Y. But the shapes of the modules are slightly different between the species, as expected if molecular coevolution has occurred.

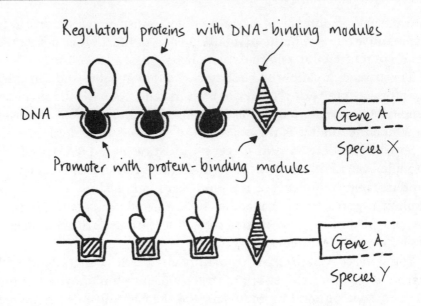

Molecular coevolution of regulatory protein and promoter binding sites *Two species, X and Y, share the same gene, A. Each gene has a promoter consisting of four copies of a binding site to a regulatory protein. However, the molecular matching between the binding site and the protein is different between the species. This is shown by slightly different shapes and shading. The regulatory protein of species X binds with too much or too little efficiency, or not at all, to the binding sites of species Y, and vice versa. For simplicity only one type of regulatory protein is shown. If species X and Y are very closely related, molecular coevolution between proteins and their binding sites might not yet have become established. Molecular coevolution is observed in other genetic systems requiring molecular binding between proteins and DNA or between proteins and proteins.*

Before I discuss how this might happen, first we need to be clear about the evidence that it actually has happened.

In several different groups of animals and plants, experiments reveal an incompatibility between the modular folds of regulatory proteins and the modular binding sites in promoters when each is isolated from a different species. For example, the promoters of the rDNA genes of *D. hydei* are not recognized by the regulatory proteins of *D. melanogaster*. The same is true of the rDNA regulatory systems of different mammalian species and of different plant species.

An analogy to explain coevolving modules is derived from special machines made to deliver canned drinks once we've put in the right amount of coins. They are similar, in principle, to the penny slot-machines at the funfairs of your day. The drinks machines used to be designed to ensure there was a different-sized slot for the insertion of different-sized coins. Each country, therefore, would have slots appropriate for the local currency. Clearly, the slots of one country would be incompatible with the coins of another. We can imagine a process of molecular coevolution happening when a country decides to change the shapes of its coins. There has to be a corresponding change in all the slots of all the machines. A turnover of coins, initiated by the banks, drives a selective modification of the slots. Does this analogy make sense?

Me and Francis Crick

Back in the early 1980s I was visiting Leslie Orgel at the Salk Institute at La Jolla in the United States. He and Francis Crick had jointly written one of the two 'selfish DNA' papers published in *Nature* that I had criticized at the time, for DNA was more 'ignorant' than 'selfish'. Calling DNA names was the sort of thing we all got up to in those days. Leslie took me to see Francis, ensconced in Jacob Bronowski's old office, where I was rash enough to say that unless we had an evolutionary perspective none of us was going to get very far in understanding DNA and developmental biology. At that point Crick jumped up and said that until we understand the nuts and bolts of biology we will never understand evolution. When he fixed me with his well-known 'now what are you going to say about that?' look, I knew my time had come. Either I said something reasonably intelligent and perceptive or I would be dismissed as another evolutionist with porridge oats for a brain. So I spilled the beans on molecular coevolution and threw the question back to my interrogators. How would they explain the seemingly never-ending joint diversification in the rDNA regulatory proteins and their binding sites? Why were the two components coevolving, one with another, in this most important of ancient genetic systems? Once I gave my evolutionary take on molecular coevolution, involving turnover, redundancy and so on, Crick had the grace to admit that it was one of the first accounts of a molecular phenomenon that could be properly understood only in the context of evolution. I started to breath again.

Today, Charles, things have changed remarkably. Bringing evolution and development together engages the minds of many biologists. But not so long ago evolutionists were quite happily ignoring what went on inside organisms – despite the fact that you included a chapter on embryology in *The Origin*. The only biology of interest was that which affected Darwinian reproductive success, so all the new discoveries were acknowledged on a need-to-know basis only. The molecular biologists were not slow in doing the opposite – in declaring evolutionary thinking, with a whiff of anti-intellectualism, as beyond the pale of real men.

Molecular coevolution is not a peculiarity of rDNA genes. It has been monitored in other genetic systems involved with, for example, the regulation of the important genes in the development of multi-cellular organisms. To get off the ground, molecular coevolution requires TRAM. I will describe why I think it is a consequence of molecular drive and natural selection, but first I want to give you a rest and let you think about the phenomenon yourself.

I am prepared to say, at this juncture, that molecular drive is pushing natural selection and not vice versa, although I would be interested to hear your views on this. In other words, natural selection could be involved as much with solving internal problems of redundant genetic systems in continual states of flux as with solving the external problems of the environment. Indeed, as I hinted earlier, we have more direct evidence for the role of selection among the molecules than we have for its more traditionally conceived role in ecology.

Selection is welcome to come on board the good ship molecular drive!

Your agent provocateur,

Dear Gabriel

I like the role for selection that you have conjured up. I'm beginning to understand now why you have been emphasizing the issue of biological tolerance from the beginning of our correspondence. You are right that the appropriate image of selection is of a lock and key, with the role of selection for ever trying to improve the fit between them. Given this, I believe we assume, perhaps far too easily, that most mutations are not beneficial. There are far more ways to damage a lock and key and prevent their interaction than there are to improve the relationship. However, with redundant genes, or with the multiple binding sites of regulatory regions, the organism can be buffered from what we might think are the detrimental effects of a new mutation. If, on top of this, homogenization is proceeding in unison, like one of your ripples in a pool sweeping through the population, then it would seem to me that the conditions are sufficiently relaxed for selection to have the time to do whatever it has to do, by way of molecular coevolution. For the rDNA promoter, this presumably means searching through the available mutant alleles of the gene coding for the regulatory protein, in the hope that one really is capable of compensating for the alteration in the protein-binding module of the promoter.

I think that this is what you imply when you say selection is busy finding solutions to problems generated by redundant sequences in the genome. Surely, you did not imply that selection, in some mysterious way, induces a compensatory mutation? I hope not, although I'm open to any possibility of what may happen in biology with your vastly improved techniques of analysis. I'm amazed to hear that you can mix the rDNA of one species with the regulatory proteins of another. Presumably this is in some sort of test-tube. Next you'll be telling me that you can transfer a gene from one species to another and get it to make the required protein inside a foreign species.

I have some further comments that you might like to receive. We should not forget that natural selection, acting alone, can encourage molecular coevolution between two molecules participating in a given function. I say this because we both agree that the phenotype is the target of selection not the so-called selfish genes. And the phenotype is the product of thousands of molecular interactions, all of which are legitimate targets of selection. Natural selection is therefore essentially the result of differences in how genes and their protein products interact. You've said as much yourself. If two genes don't get along with each other, the individual bearing the squabbling genes suffers the consequences. It would be wrong to imply that molecular coevolution is unique to molecular drive, pushing selection to promote compensatory mutations. Or do you see 'molecular coevolution' as something different and special?

On the issue of internal buffering, how can one measure this? Surely, the more genes, the more effective the buffering. Five hundred copies of the rDNA gene provide a convincing enough argument that a few mutant repeats are going to be largely inconsequential to the organism. But what if the redundancy gets smaller, down to the extreme of only two copies of a particular gene? Are there such minute families of genes and how do they evolve? Is molecular drive at all pertinent in such cases?

On another point, I can understand how concerted evolution might exist as a way of discriminating genetically between species. For example, species A might have a gene family homogenized for mutation X, whereas species B has been homogenized for mutation Y. Indeed, is it not possible that the patterns of concerted evolution are precisely 'mappable' (if I may use that phrase) to species? In other words, are your processes contributing to the origin of new species, or are species the natural barriers preventing further homogenization? This could be a chicken-and-egg problem, but I would like your views on this. As you know, I could never get around this problem of the generation of new species with natural selection alone. How could selection promote sterility of the offspring between individuals on the first steps towards being separate species? This goes against the very operational definition of natural selection. There is a peculiar paradox to the origin of biological barriers between species. I wrote to Wallace about this who wrote back with nineteen mathematical theorems trying to solve the problem! I recall telling him that his paper made my stomach feel as if it had been placed in a vice and drove three of my children half mad.

Molecular drive could be involved in the initiation of new species if the additional phenomenon of molecular coevolution could be shown to be a generalized way of biologically distinguishing one species from another. You say that the proof of molecular coevolution is the failure of molecules to make proper functional contact with one another if the two molecules come from different species. You are observing a biological incompatibility between species for a given function normally carried out by consenting molecules in each species. If you could show that molecular coevolution and the observed incompatibility between molecules from different sources always exist between closely related species, you could have a mechanism for the evolution of species on your hands. What do you think of this possibility? I'm sure I don't know all the facts yet to be able to decide for myself.

If molecular coevolution is, as you say, the product of molecular drive among TRAM systems and natural selection among other genes that functionally interact with the TRAM systems, and if this turns out to be a mechanism for the successful origin of new species, then I'm more than happy that natural selection has a major role to play in all of this.

Your stimulated correspondent

THE MYSTERY OF MYSTERIES

Dear Charles

There is a war raging in the heart of Europe. Complete communities of supposedly genetically and culturally distinct populations are being massacred by neighbouring communities. Pitiless scenes of human misery are on view every day, every night. Where does such hatred come from? This current war in the Balkans is only the latest in a series of worldwide wars that define the twentieth century. The so-called Great War (1914–18) decimated the male population of Europe. The Second World War (1939–45) systematically murdered millions of 'undesirables' – mainly Jews – and culminated in the physical obliteration of some of Europe and Japan's great cities. And now, again, we're seeing the horrors of 'ethnic obliteration', taking place only a few hours' flying time from where I sit in the comfort of my Oxford home, writing letters to you on the tolerant, cooperative ways in which biological organisms seem to have evolved. Maybe life is 'red in tooth and claw' after all – or at least that particular form of life known as the 'human being'. And the tragedy of these racial wars is that they are seemingly rooted in the triumphalist ideology of 'natural selection' in ways that would have you shivering in your tomb, if I gave you the details. For many people, human culture and civilization stopped at the doors of the German-inspired gas chambers, where six million 'selections' took place: you die, you die, you don't, you die, you don't.

Is this base human nature part of the grand scheme of things – the pinnacle of evolutionary achievement at the end of the furthermost twig on the longest branch of the tree of life, accessible to only the most cunning of our genes, as they scratch and fight to survive? Is such inhuman behaviour one of the 'universals' of our human condition, determined by our genes as they survived countless battles in the murky evolutionary past? Are such basic, murderous instincts capable of being aroused in you and in me as we exercise ourselves with the scientific understanding of life, and not with the scientific obliteration of life? Would we all become 'animals', if placed in the appropriate circumstances?

You must forgive me, Charles, for starting this letter in this way, but the horror of the daily massacres makes it difficult to function in my ivory tower. I can only hope that one day the futility of finding a genetic basis for the discrimination of human 'races' will be clear to an enlightened world. There is a lot of genetic difference between individual humans, but this does not break down along ethnic lines. The vast majority of variation exists within groups of humans, not between them. Indeed, most of the variation is shared between groups. If you took an individual, measured some characteristic such as intelligence, height, humour and so on, and tried to decide what group he or she belonged to, you would no more get it right than by tossing a coin in the air. We are all genetically unique and we are all genetically human. The individual and the species are the only two units with recognizable biological boundaries. Everything else – ethnic groups, populations, races, subspecies and so on – are largely unmeasurable abstractions.

An origin of species?

I was half expecting your latest question about whether molecularly driven changes in TRAM systems could generate new species. As you yourself wrote, the formation of species is the 'mystery of mysteries', and that is still very much the case today. There are many different definitions of species, but I prefer this commonsensical one. Individuals of a species are sexually and developmentally compatible with each other, but they usually do not mate with individuals of another species, or if they do, they produce inviable or infertile offspring. This definition has the merit of emphasizing biological compatibility within a species and subsequent incompatibility between species.

I will tell you a story of genes in the fruitfly *Drosophila* that have all the features of TRAM and that could be involved directly with the evolution of species. This would imply a role for molecular drive, along with selection, in the establishment of new forms and behaviours.

Love songs and species

This story relates to a particular gene called *period*, involved with biological rhythms in our old friend, the banana fruitfly. It contributes, first, to the male fly's ability to produce rhythmic love songs; second, to the perception of the 24-hour day and night cycle; and third, to the overall developmental cycle of several days as the egg matures into an adult.

The time level most relevant to our discussion of the development of species is the production of the male's love song. This is achieved by a rapid beating of the wings of the male, which sends out an audible rhythmic sound detectable by a nearby female. If she isn't put off by the sound, he is allowed to do a little dance, lick her genitalia and eventually mount. The love-song signal and the courtship display are specific to each species. For example, tape recordings of the wing beats of *D. melanogaster* and *D. simulans* show that the beats oscillate at very different frequencies. Indeed, enough is known about love songs of *Drosophila* to be able to predict that each of the five thousand species probably has a different way of doing the same thing. That is, males are most successful in attracting females of the same species. So how is this achieved?

Despite other mating signals involving smells and behaviour, it seems that the *period* gene's control of male wing-beat patterns plays a role in convincing the female that the male is one of her own. Females of a given species can be fooled into thinking that they are being courted by males of a different species by cutting off the wings of their own males and playing the tape recording of the love songs of males from another species.

A key evolutionary question is: how does any new love song get off the ground and spread itself around? A mutant male with an unusual love song is not going to be very easily accepted by the females, who would naturally consider themselves safer in the long run with males beating out more familiar songs. To answer this type of awkward question, we need to do a little molecular dissection of the *period* gene.

Indeed, Mr Darwin, this part of the investigation has entailed the moving of the *period* gene, or parts of the gene, from one species to another in the way you predicted in one of your earlier letters. I'm sure that you do not want to hear how precisely this is done. It is a very long experimental process starting with the first problem of tracking down the gene, through to its eventual cloning and sequencing, and on to the triumphant comparison of how it works, both in the species from which it was taken and in the foreign environment of another species.

As with thousands of other genes, the *period* gene is partly internally repetitive, in that some subsection of the gene is repeated over and over again. In *period* the repeats are in a tandem array like the railway carriages I described in an earlier correspondence. The length of the repeat is only six bases long and codes for two amino acids, threonine (T) and glycine (G). The shortness of the repeat probably means that they arose by the slippage mechanism of turnover. The co-existence of alternative lengths of *period*, each having a different number of TG repeats, indicates that slippage is still going on in *D. melanogaster*.

So we have a tandem array of TG repeats embedded inside a gene that can vary in the number of repeats, as trains can vary in the number of carriages. Do the TG repeats, or the fluctuations in numbers of TG repeats, have any influence on the *period* gene's contribution to male songs and sleep–wake cycles?

Concerted evolution in the *period* gene

The first thing that stands out from a dozen comparisons of the structure of *period* between species is that there is indeed a repetitive region in each species and that this region often contains repeats of a completely different length and composition. In some species the repeats can consist of a string of five amino acids arranged in an array about four times longer than the array in *D. melanogaster*.

The comparison of *period* between species reveals the existence of concerted evolution. We can therefore identify a species by the composition of its *period* repetitive region.

Period turnover

What lies behind this concerted evolution and does the repetitive region affect the biological function of the protein encoded by the *period* gene?

The arrangements of sequences in the *period* gene's repetitive region reveal that in addition to slippage, unequal crossing-over and gene conversion are occurring. Both these mechanisms can operate quite happily between chromosomes, so they could readily promote the spread of new repetitive patterns and sequences through a sexual population, as a molecularly driven process.

Molecular coevolution in *period* genes

Has the molecularly driven establishment of concerted evolution affected the function of the *period* gene? To examine this we need to look for evidence of molecular coevolution, the product of the interaction of natural selection and molecular drive. For this we need to move genes, or parts of genes, from one species to another.

The first hints of molecular coevolution in the *period* genetic system came from comparisons of the structure of the *period* gene in the dozen different *Drosophila* species that I mentioned earlier. Remember that the repetitive region can vary between species because both the sequence composition and length of the repeats have changed over time. But we should not forget that there are another 4,990 species of *Drosophila* to be examined!

If we had a map of the gene on the page in front of us, we would see that the repetitive region divides the gene into three parts. The region to the left of the repetitive region is far more variable than the right-hand region. There is an unexpected association between specific mutations in the left-hand flanking region and variation in the composition and length of the repetitive region. Does this association have a functional basis? This can be answered by creating novel mosaic genes using parts from two different species. The idea now is to mix-and-match the repetitive region of one species with the left-hand flanking region of another species. Can they function successfully together? One test is to inject the eggs of a mutant *D. melanogaster* that has lost all its rhythms with different *period* gene mosaic constructs. If the rhythms are restored, the mosaic construct has normal *period* functions.

When the repetitive region of *D. pseudoobscura* is used to replace the repetitive region of *D. melanogaster*, there is no restoration of rhythm in the injected mutant fly. But if a *D. pseudoobscura period* gene consisting of its own repetitive region plus its own left-hand flank is injected, rhythms are restored. So the two regions, derived from the same species, have to lie together to function successfully. The two regions have coevolved as a pair.

Before I go any further I have to come clean about one thing. I've rather casually referred to restoration of rhythms in these mix-and-match experiments without saying which rhythm I am talking about. The experiments were done using the sleep–wake cycles as a functional test, and not the male love songs, because the sleep–wake cycles are much easier to measure. But there is no doubt that the two rhythmic cycles of sleep–wake and male love-song beats are influenced by the same *period* gene.

The *period* gene works successfully only if the appropriate repetitive region is surrounded by the appropriate coevolved left-hand flanking sequence. Any other combinations, derived from different species, do not restore rhythm in the mutant fly. We can hypothesize that the changes in the fast-evolving repetitive regions, due to non-Mendelian turnover, have caused compensatory changes in the flanking regions. The former are spread by molecular drive and the latter are spread by natural selection. Together they ensure that, throughout a period of change in biological rhythms, there is no complete shutdown to an unrhythmical and presumably sterile state in this important gene. Molecular coevolution can be seen to have saved the day, once again.

By using a combination of molecular drive and natural selection, it is possible to lessen the problem of how a novel male love song spreads through a population. The homogenization of the repeats in the *period* gene by a variant repeat would not have been rapid enough, leaving a male with a song unloved by resident females. Given known rates of turnover, the replacement process would have been slow and gradual. Furthermore, the population would have changed in unison, for the reasons I have explained in earlier letters. In any given generation, any change in rhythm would therefore more or less embrace all individuals living in that generation. This provides the relaxed conditions for selection to promote the observed molecular coevolution not only between the repetitive region and the left-hand flanking region, but perhaps also between the slowly accumulating new love song and receptivity of the females to the new situation. Nothing is known so far about any such coevolution between the sexes, but it must happen if males and females of a given species are to turn each other on successfully.

I'll leave this story at this point, to let you get some rest. I look forward to receiving your comments in due course.

As ever

Westminster Abbey

My dear Dover

Your last letter about period *and species formation is intriguingly suggestive, although its opening remarks on the bleakness of human racial conflict is depressing. We had our wars too, but nothing like the scale you have described to me. I have been genuinely shaken and shocked to hear that some of these irrational conflicts have tried to seek some sort of moral and theoretical justification from my theory of natural selection. Surely, the speed at which human society develops on an historical timescale far exceeds what could be accomplished by way of change on an evolutionary timescale. As you know, I do believe that some aspects of human nature are governed by our evolutionary past, in the same way that individual development reflects the long march of evolution, right back to the origin of life. I recall you saying in your very first letter that you were somewhat against this idea of a biological underpinning to human nature – as it is expressed in a discipline you called 'evolutionary psychology'. You haven't yet expanded on that and I'm keen* to hear more from you on this subject, for I would find it difficult to accept that there was any aspect of humans that had evolved in ways that are different from those affecting all other living forms.

Acceptance of a biological basis to human nature should not mean, however, that we have no free will in how to behave. The availability of choices surely depends on how humans have organized themselves into societies, an activity that can be passed on far more easily through example and by word of mouth than by the genes. I say this because it is relevant, I believe, to the issue of human races. There were a lot of people in my time who were convinced that races could be ordered into a hierarchy of types from the lowest (inferior) to the highest (superior). I do not need to spell out who is supposed to be at which end. I am guilty myself of such ideas, as you'll see in my earlier writings.

Although there are clear differences between races in facial features, general body shape and skin colour, it would be wrong to extrapolate from these a scale of evolutionary advancement. We simply do not have a measure of the direction of time's arrow. For example, did 'white' skin pigmentation evolve from 'black' skin pigmentation? Did Negroid facial features evolve from Caucasian facial features? Maybe you will tell me that you have some answers to these questions. But, for the moment, we cannot say, even if we considered this an important question, which races are biologically 'superior' and which are biologically 'inferior', or, to be more exact, which are the ancestral races and which are the derived ones. My suspicion is that all human 'races' are equally old relative to the much greater distance that separates us as a species from our nearest chimp relatives.

The race issue is a morass of confusion from several other perspectives too. Where does one race end and another begin? Which criteria do we use to measure, distinguish and catalogue races? Morphology? Molecules? Behaviour? Intelligence? I know that my cousin Francis Galton was keen on establishing a science of such measurements with a view to human improvement through eugenics, but I do not think that any comfort could be derived from natural selection about the feasibility of such objectives. You see, Dover, natural selection is a slow business, taking place over hundreds of thousands of generations. The very slowness of the process *begs two important questions. Has there actually been enough time, since the emergence of* Homo sapiens *as a separate species, for natural selection to produce racial differences that would justify either a hierarchy of evolutionary advancement or the appearance of races with real and observable biological differences in form and behaviour? Second, how can the selective breeding of humans (for that is what it would take) expect to achieve, in historical times, what could be achievable only on an evolutionary timescale? Surely, no one is seriously proposing that all 'undesirable' humans are stopped from breeding, while the selected few get on with it. I feel that such human manipulation in the name of inheritance and biology is a serious distortion of the mechanism of natural selection, which leaves me quite worried about what is being said in my name.*

I agree with you that 'race' or 'ethnic group' is not a category with much biological meaning. As you said, individuals have a clearly defined boundary and mode of organization (give or take some clonal plants and animal colonies), as do species (although there are ambiguities about species boundaries in some cases, particularly plants). But race is not a well-defined category, and certainly not sufficiently clear at your level of genetic dissection to justify eugenics or 'ethnic obliteration'. I would like to hear more from you on this. Presumably, you are much further on in the study of the genetic basis (if any) of different human populations. I suspect that the issue of differences in psychology and behaviour between men and women could also be as contentious as the issue of intelligence between races, or even between our English social classes.

I am sure you could pick through many of my early writings and find evidence that I too, like many of my Victorian contemporaries, was happy to place races on a scale of 'inferior' to 'superior'. Indeed, I would understand it if some of the things I had to say about race were considered overt support for the idea that Negroes or native Australians were nearer in biology to some of the great apes than to the Caucasian.

Similarly, re-reading my own words, I might not be too proud of the way I depicted the inevitable accumulation of biological differences between men and women as the result of sexual selection –

that is, the selection of characteristics peculiar to one sex (for example, male peacock's tails) as the combined result of sexual competition between males and the sexual preference of females for well-endowed males. Social conditioning too must be a very powerful force shaping sexual attitudes and aspirations. Luckily for me, I also wrote that racial differences in social grace and behaviour are largely conditioned by the harshness of the surroundings. Even in my worst descriptions of the deplorable state of the Fuegians living at the southernmost tip of South America, I considered them to be biologically identical 'fellow creatures'. My attitude towards 'inferior' races was simply that they were 'uncivilized' when judged by the definitions and standards of civilization of white Europeans. My overriding feeling was that all such unfortunates could be 'improved' if placed in the right conditions and under the appropriate Christian direction. I have to say that I am proud to have taken an active part in the anti-slavery movement. I am quite indignant about the use of what is called 'Darwinism' to categorize races into fixed and immutable units, for ever shackled with their selected biological differences. If differences in race or sex do exist, we should make it clear that improvements in living conditions are more likely to minimize such differences and bring so-called 'inferior' races up to standards of 'superior' races than is short-term eugenic manipulation of human biology.

Let me try to make a connection between what I've just been discussing about races and our current discussion of the origin of new species. A reader might believe by mistake that I am prepared to give credence to the biological entities of individuals and species, but not to the units in between, such as races. I admit that there might appear to be some contradiction here. After all, I did give my book the alternative subtitle 'The Preservation of Favoured Races in the Struggle for Life'. To clarify my position, I am against the view, first ascribed to Plato, that species have fixed identities for which one 'type' specimen can be found and preserved in a museum. This 'typological' way of thinking goes against everything I proposed about evolution. Evolution is based primarily on differences (some of them inherited) between the individuals in a population. Such fine-grained differences are the very stuff of evolution. There can be no 'type' specimen if evolution is to proceed. Species are not abstractions; they consist of many biologically similar individuals capable of reproducing with each other, generation after generation.

As an evolutionist, I focused hard on individual differences becoming transformed, in the fullness of time, into species. However, as with all processes, the stages of transition might be observable as slightly incompatible groups of organisms below the level of species. 'Races' (or whatever word we wish to apply) could be one of several transition stages. But the short timescale of human evolution coupled to the rapidity of the cultural spread of human improvement means that our so-called human 'races' could be the result as much of differences in environment as of differences in biology. There are some biological differences between races, as I've said, but these could be irrelevant when we are considering such flexible characteristics as human behaviour and culture. Surely, Dover, something more can be said about these things, using your molecular techniques of analysis. As you can see, I'm getting quite impatient and irritated about what I'm supposed to have said myself and the uses and abuses of natural and sexual selection.

Is *period* a general or special case?

The outcome of evolution in your period *gene example has meant a change in a pre-mating barrier, as seen in the male's love song. But surely there are, or should be, lots of examples where changes in external appearance too have been the direct outcome of evolution. After all, for most people in my day, and I should imagine that it is the same in your day, the differences in appearance of species are clear and obvious. Cats differ from dogs; oaks from beech. People are not familiar with many of your* Drosophila *sibling species, which to all intents and purposes look alike. But among different species of birds or frogs, even with extensive differences in mating calls, the species in question often look very different from each other.* •

How did all these obvious differences in shape arise between species? Did they accidentally accumulate after an initial species-forming event involving, say, something like the period *gene or were they a cause of species formation? What do you know of the genes involved with body shape? Are they TRAM systems, similar to the genetic systems you've chosen to describe to me so far?*

The period *gene example suggests that the distinction between pre-mating and post-mating barriers between species is a little artificial because they have equal status. It had been thought that pre-mating barriers would be reinforced by selection to stop wastage in the production of inviable or sterile hybrids. Perhaps changes in the* period *gene just happened to affect male love songs – a pre-mating event. In a similar vein, there may be genetic systems with TRAM characteristics that just happen to produce post-mating barriers. In other words, we are not in any position to say that pre-mating barriers arise from post-mating barriers. Either one can happen first, presumably depending on whatever genetic system is affected by molecular drive and natural selection.*

Your most sincere friend

Charles

BIOLOGICAL BARRIERS

Dear Charles

The misappropriation of natural selection in the definition of individuality

I can well understand your frustration with the misuse of natural selection in human affairs. I tried to show why this might be the case in my first letter when I restored the individual to centre stage in place of the supposedly all-powerful gene. But no matter how misunderstood selection might be, and no matter how many different definitions of natural selection one can find in textbooks and professional publications, there is no excuse for the worst excesses that are carried out in your name. Self-justifying explanations for the dominant role of deterministic genes in the formation of race, class, gender, sexual orientation, social behaviour and cultural mores sprinkle many fashionable attempts to dissect the human condition. We are our genes, it is thought.

One recent innocent example springs to mind that will give you a flavour of the problem. One of England's powerful literary voices, the poet laureate Ted Hughes, died a few months ago. A eulogy at his funeral was movingly composed by his great friend and comrade-in-arms Seamus Heaney, an internationally acclaimed poet who had been awarded the Nobel Prize. The eulogy was sincere and genuinely original in trying to analyse the makings of Ted Hughes. Unfortunately, for me as a geneticist, it unwittingly bordered on the incoherent as it paid lip service to what Heaney considered to be modern thinking about our genetic heritage and destiny.

Heaney remarked 'One part of Ted Hughes believed in the gene and its laws as the reality we inhabit ... from alpha at the start of the evolutionary journey to omega at the end. But another part of him looked into the visionary crystal, and could see Dante's eternal Margherita, the pearl of foreverness, in the interstices of DNA.'

For me, as a life-long addict of Hughes's poetry, the black grin of the 'crow' at the very sick centre of God's creation of man, as he tries to accept the obdurate forces of sex, self-consciousness and death, underscores the originality of Hughes. If genetics is to have played a part in this originality then, like every other aspect of individual phenotype, we can firmly say that such originality never occurred in the past nor will ever occur again in the future. I'm happy to accept that the random sample of genes inherited by Hughes from his mother and father played no small measure in the development of a receptive and creative mind. But such genetic influences, even without any environmental input (a highly unlikely situation), cannot be described as Hughes's link with the deepest forces of our evolutionary past, leaving to his particular individual insight the simple role of filling in the gaps between the double helix.

Hughes, like every one of us, is nothing but pure, unadulterated 'individual'. There is no biological connection with the past or with the future. We uniquely live only in the present, even though our ignorant genes were inherited from generations past. What cannot be inherited is the particular collection of interactions of genes with genes and of genes with environment that uniquely defines each of us. Each gene is a unit of inheritance, but it can not carry into the next generation the unique set of interactive functions that defines every individual. A gene cannot control which genetic partners it finds itself with in any individual. The one-off set of genetic interactions that specify the unique unfolding of an individual cannot be inherited. A result of this, as I argued extensively in an earlier letter, is that the genes, in themselves, cannot be the targets of selection, nor can they be, individually and independently, the arbiters of our creativity. Some of Bach's twenty children were composers but many were not and none was like the father. As for the grandchildren ...

I will not delay too long, Charles, in answering your concerns about the genetic basis of race. But I cannot really answer your questions in detail, as befits their nature, until I've described the modular way in which biological organisms construct themselves. You will appreciate, I hope, that this has an intimate bearing on what we can and cannot say about the genetic basis of differences between populations, regarding such controversial issues of intelligence, educability, motivation and other issues that govern our societies today.

I think it best at this juncture to come back to your more direct questions about the *period* gene.

How many roads must a species travel?

Your questions raise the fundamental issue of the many different roads leading to the formation of species. As with everything else in evolution, there are many aspects of biology that can go 'wrong'. It is notoriously difficult to get at the genes involved with reproductive isolation, whether these arise before, during or after mating. But some interesting progress has been made.

We can now exploit experimentally the fact that the hybrid offspring between some species of *Drosophila* are not always inviable or sterile. Curiously, although the male hybrids are sterile, the females are fertile. This difference means that females can be used for further genetic crossing in experiments designed to isolate the genes involved with male infertility. Two of the species most amenable to such manipulation are *D. simulans* and *D. mauritiana*. They are almost indistinguishable in body shape, except for the male genitalia. *D. simulans* is cosmopolitan, having learned to live successfully on human garbage and fruit products. *D. mauritiana*, as its name implies, is restricted to one island, Mauritius, isolated from surrounding islands for the past half-a-million years. The hybrid males of a cross between the two species have defective sperm and so are sterile, in contrast to the hybrid females, which retain normal fertility.

After years of painstaking genetic crosses between the species, no fewer than 120 genes have been found that are responsible for hybrid male sterility. They are scattered all over the chromosomes, including the sex chromosomes, of both species. Each gene has a small effect on sterility; nevertheless they strongly interact with each other to bring about their effects. In other words, the genes (or more accurately their protein products) are collectively engaged in a number of key biological functions involved with sterility. So once again we see that important functions arise from the combined effects of interacting genes. No gene is an island unto itself.

The reason I'm stressing the interactive basis of the genes is that it is quite possible that molecular coevolution has occurred between functionally intertwined genes. In each species the way in which the genes interact in achieving fertility may be different. If so, biological incompatibility will arise (in this case, sterility) when the genes, taken from two different species, are artificially mixed in the hybrid.

A strong hint that this is indeed the case comes from success in finding genes that can abolish hybrid male sterility normally observed in crosses between two other *Drosophila* species, *D. arizonae* and *D. mojavensis*. Normally, hybrid sterility between the two species is caused by genes on the Y chromosome of *D. arizonae*. But sterility can be abolished when a specific region of chromosome 4 of *D. arizonae* is also present.

There must be a gene (or genes) on chromosome 4 of *D. arizonae* that interacts with a gene (or genes) on the Y chromosome for normal male fertility. In *D. arizonae*, this particular interaction is successful. However, between the Y-based gene of *D. arizonae* and the chromosome 4-based gene of *D. mojavensis*, the interaction is unsuccessful. Fertility is resolved only when chromosome 4 of *D. arizonae* is present. It is possible that the product of one of the pair of interacting genes is a regulatory protein that interacts with binding sites in the promoter of a gene involved with fertility.

Odysseus leads the way

In the case of *D. arizonae* and *D. mojavensis*, nothing more is known about this situation. But one of the 120 interactive genes responsible for male fertility in *D. simulans* and *D. mauritiana* turns out to encode a regulatory protein that can bind to the appropriate sites in the promoters of other genes. This gene has been called *Odysseus* for reasons that I will not go into. Fruitfly geneticists are generally endowed with a good sense of humour and a catholic education from classical to pop. As such, genes are given all sorts of weird and wonderful names.

Odysseus produces a 'homeodomain-bearing transcription factor'.

A what? you may ask. This is genetical jargon for a protein (a modular part of which is called the homeodomain) that can regulate another gene's activity by binding to specific sites in the promoter of the other gene.

The homeodomain of *Odysseus* probably binds to target sites of genes involved with male fertility. Surprisingly, the *Odysseus* homeodomain sequences of four closely related species of *Drosophila*, including *simulans* and *mauritiana* differ in fifteen out of sixty amino acids. Usually, homeodomains of different proteins share pretty much the same sequence of amino acids, even those from different species.

So perhaps the evolutionary forces responsible for species formation, particularly the genes involved with male sperm development and male genitalia, are responsible for the considerable divergence in the homeodomain module of the *Odysseus* regulatory gene. These forces could involve non-Mendelian turnover mechanisms busying themselves among the multiple binding sites targeted by the *Odysseus* homeodomain. Although the target genes and their promoters are not known, there is evidence that they exist because male hybrid sterility can be reduced when other genes are cotransferred with *Odysseus* from one species to another. Presumably, these are the genes whose promoters are regulated by *Odysseus*. What we are observing is a biological incompatibility between species, probably due to molecular coevolution between *Odysseus* and its target promoter binding sites in the genes involved with male fertility.

I am not suggesting that this example of yet another TRAM genetic system is solely responsible for differentiation among these species. We should not forget, for example, the argument you first made about the effects of sexual selection as a driving force in species formation. Competition between males for mates, coupled to female preferences for certain desirable male features, could very well influence the most sensitive aspect of a male's reproductive biology – his penis and his sperm. A potent mixture of sexual selection, natural selection and molecular drive could lead willy-nilly to rapid genetic isolation of one species from another.

So let me conclude our discussion of the 'mystery of mysteries' with two further recent discoveries involving vertebrates. The first concerns two mammalian species, the rat and the mouse. A homeodomain-containing gene (called *Pem*) has been found to be active in their testes. This time there are an astonishing twenty-four amino acid differences in the homeodomain between these two relatively closely related rodent species. I'm convinced that it will not be too long before we have evidence that *Pem* is involved with sperm production and that molecular coevolution contributes to the biological infertility of hybrids of these two rodent species.

My second example concerns fish. I'm sure you have remained fascinated and bewildered to this day by the spectacular variety of ray-finned fish living in the coral reefs that you studied on your *Beagle* voyage. They include every possible body shape: goldfish, seahorse, flounders, pufferfish, angler fish and butterfly fish. All in all, there are more than twenty-five thousand different ray-finned species, much more spectacularly divergent in appearance than the boring species of *Drosophila*.

Where has all this diversity come from? What is driving the underlying genetic systems to continually differentiate from each other?

One of the most intriguing answers to this conundrum could arise from the study of fish *Hox* genes. These are the 'master' control genes involved with important decisions of construction during individual development in most species of animals. I will be having a lot to say about *Hox* genes in due course. To everyone's surprise, ray-finned fish can contain up to seven clusters of *Hox* genes, with the potential to have up to thirteen *Hox* genes in each cluster. We humans have only four such clusters and for some time four has been considered the pinnacle of evolutionary achievement. In primitive animals, such as insects and worms, there is only one cluster. Yet here we have seven clusters of *Hox* associated with an explosion of species formation, leading to more than twenty-five thousand different types of body shape (presumably together with twenty-five thousand different behaviours). Is there a connection between these two events?

As ever

Gabriel

My dear Dover   Westminster Abbey

    Your last examples are beginning to show that there is a connection between rapid rates of divergence in species morphology and the existence of genetically based barriers affecting pre-mating or post-mating reproductive barriers. I suspect that you have a lot more to tell me about the genes affecting morphology and behaviour, but from what I've heard so far I can understand why they interest you. They underline your point that differences in how genes interact with each other seem to be the distinguishing features of biological evolution; and that the genes involved are typically TRAM systems. As such, their involvement in the establishment of new species has been a mixture of natural selection and molecular drive. I suspect though that, until you tell me more about the genes involved with individual development, I cannot fully assess the magnitude of any of these systems. I really am quite eager to hear more about the Hox 'master' genes. Could such genes be involved with the 'punctuated equilibrium' idea of Stephen Jay Gould and Niles Eldredge in generating bursts of new body shapes followed by a long period of inactivity?

    But before you enter into Hox, could you indulge me in satisfying my curiosity about a remark you let slip about sex. You seemed to imply that sex has an important role in species formation in that its main task has something to do with keeping all the members of a species together. This seems to contradict the view that sex is a means for generating differences between individuals, a point you have strongly emphasized yourself in your letters about the uniqueness of individuals in every generation.

    Before we leave the 'mystery of mysteries', I'm intrigued to hear more about your comment on sex, unless it was just a casual off-the-cuff remark.

    Your obliging servant

SEX — A NEW PERSPECTIVE

My Dear Charles

Yes, of course I will indulge you in a little more on species formation before we move on. My remark about sex was not casual and I'm glad you've given me the opportunity to expand on it. The idea does in a sense encapsulate what I perceive to be the essential link between the pattern of unfolding of a fertilized egg into an adult and the evolution of new patterns of unfolding, recognized as different species. Sex plays an important role in this link between development and evolution.

The best way to start is to go back to your question of whether concerted evolution (remember the families of redundant genes that have been homogenized for different mutations in separate species) is connected in any way with species formation.

When I first started researching into concerted evolution, I thought, like you, that there was an uncanny pattern to the homogenization of mutant repeats, with each separately homogenized family of repeats being unique to a species. Was concerted evolution trying to tell us something about the formation of species?

Strictly speaking, no. Concerted evolution is not a watertight species-distinguishing genetic phenomenon. And redundant genetic elements come in all shapes and sizes in a wide variety of genera. So the underlying genetic process of molecular drive is no more and no less a process for the formation of species than is natural selection or neutral drift. Molecular drive can contribute to the origin of biological incompatibilities between species in the same way that natural selection may be involved. But for neither process can we say that the formation of new species is inevitable. There can be no one single feature of biology, be it morphology, development, metabolism, physiology or behaviour, which can be said to be the 'factor' for species formation. Organisms differ not so much in their component parts as in the ways in which such shared parts interact with each other, at all different levels of organization from the genes to the whole organism.

'Molecular coevolution' gives us strong hints about how biological incompatibilities might arise through an interaction between natural selection and molecular drive. Yet I'm not prepared to say, hand on heart, that even this intriguing phenomenon is a mechanism of species formation. Are the evolutionary changes in the *period* gene, which show all the features of TRAM systems required for concerted evolution and molecular coevolution, the cause or consequence of species formation? We do not know at this stage.

What's more, large families of repeats will take a long time to be fully homogenized, so much so that one wave of homogenization can be followed by another – like ripples in a pool, as I explained earlier. The number of ripples in the pool and the size of the pool will depend not only on the internal dynamics of turnover in a given family of sequences, but also on the sexual dynamics, breeding behaviour and general ecology of the organisms bearing the gene families. In these circumstances, it is not possible to say with which ripple a new population of organisms has arisen that is no longer compatible with other, differently homogenized populations.

Additionally, there are thousands of different pools of all different sizes, each representing a different family of repeats, all with their ripples. Should we define species by sequences involved with male love songs? Or should we define them through the known incompatibility of rDNA regulatory elements, to take but two examples? The problem in answering such questions is that there are no general rules for species formation: either or both of the above examples may be true in a particular set of events. Sex too has a role in much of this.

The basis of stasis

There are two complementary features of Gould and Eldredge's 'punctuated equilibrium' idea. New forms of life burst on the scene rapidly compared with the much longer periods when nothing much seems to happen. I'm not going to give you examples of 'punctuated equilibrium' because you'll find these adequately covered in Gould and Eldredge's individual books, which are very well worth reading.

From a genetics point of view, the periods of seeming immutability (stasis) are the more intriguing. The timespans during which 'bursts' of new species arise are well within the time required for species differences to accumulate gradually by any of the available procedures: natural selection, neutral drift and molecular drive. Stasis, however, seems odd given the huge timespans involved – sometimes over millions of years, covering large changes in climate and environment. With evolution by natural selection we would expect to see, at the very least, some continual selection of new adaptive forms as ecological pressures change. The argument that selection is busy eliminating all new variants in order to stick with the status quo is a little desperate given the variety of circumstances under which stasis continues to hold.

There is a molecular drive explanation, which I'll give to you here, for what it's worth. As I've emphasized so far, most genetic systems have TRAM features, including those involved with body shape and behaviour. I've therefore described how molecular drive contributes to the establishment of biological novelty. Does this mean that in periods of stasis molecular drive has shut itself down? The answer is no, even though turnover mechanisms can stop and start in different regions of the genome for entirely mysterious reasons. In general, we have to conclude from the evolutionary comparison of genes and genomes that turnover does not cease. So the spread of novel genetic arrangements in genes or in their promoters does not cease. However, there is no reason to expect that every molecularly driven novelty at the DNA level will also manifest itself at the level of physical form. Many DNA alterations can be neutral in their effects on phenotypes. Only very rarely might a novel feature spread by molecular drive substantially contribute to a phenotypic change that leads to a new species. In this case, the novelty hits the jackpot! I say 'only very rarely'. This might sound sloppy and uninformative, but it is the best that can be said for biology. As I've said above, we simply do not know what changes lead to species formation; accordingly, it would be rather pointless to quantify such unknown procedures.

Turnover and molecular drive are rather like banknotes. There is a continual turnover of new for old banknotes, operated by the banks.

Minor design features are spread quickly through the population of notes without anyone noticing. Similarly, if the central bank wants to introduce a major new design it uses the existing system of turnover. It does not recall all old designs on day 1 and issue all new designs on day 2. But the rarity of a major new design means that it is noticed. It looks like each new design is followed by a long period of stasis when nothing happens, but this isn't the case. Molecular drive never stops, although it isn't always doing something interesting.

Sex makes the world go round

You were quite right in drawing attention to natural selection as a means for the origin of new species, even if you did not tell us very much about it in your book of the same name. If the establishment of novel adaptations leads to biological incompatibilities, new species arise. But there is no inevitability in this. Biological evolution is a bizarre business. The same is true for molecular drive – some homogenized repeats will not affect biological functions, whereas some will. Of those that do, some will incite selection to promote the coevolution of interacting genes, while some will not. When molecular coevolution is established, it might not exist between closely related species, but only between distantly related species, or further apart between separate genera.

What is important about molecular coevolution is that it signifies a way of tolerating molecularly driven changes in TRAM systems without destroying essential biological functions. As I've said earlier, it is akin to changing a plane's design while the plane is flying in the air. At the heart of this essential biological trick is the ability of interacting molecular components to remain compatible with each other in order to bring about a certain biological activity. And it is the combination of the peculiar dynamics of molecular drive, coupled to sex, that allows individuals to remain biologically compatible throughout a period of change. If we think this one through, sex and species formation are therefore intimately bound up.

Sex for me is a means by which biological tolerance is maintained between the two new sets of parental chromosomes that go into the formation of each new offspring. It is crucial for the coherent developmental unfolding of each new multicellular individual that the fertilized egg receives compatible genetic 'instructions' from each of the parental sets of chromosomes. Everyone needs to sing from the same hymn sheet.

It might not matter, in an absolute sense, what these precise 'instructions' might be, provided that they are giving out the same message. And it would be important for the survival and reproduction of the whole population of newly developing individuals that, on average, the parental 'instructions' are compatible in each and every fertilized egg. Again, the absolute nature of the 'instructions' might not be important. Rather, it would be of more importance for the long-term health of a population to prevent the build-up of any big differences in developmental 'instructions' between any two sets of parental genes, as they randomly meet each other.

It is the combination of chromosome shuffling through sex and the spread of genetic variants between chromosomes by turnover that ensures a continual compatibility between parental genomes even during periods of evolutionary change. Essentially, sex ensures that once turnover has homogenized a given variant between two chromosomes in one of the parents, in the next generation homogenization will inevitably involve two different chromosomes. In this way, all chromosomes are kept in touch with each other and acquire any new 'message'.

In addition, sex and molecular drive ensure a slow cohesive change in populations, making it easy for selection to help to establish molecular coevolution between functionally consenting molecules.

What I am proposing is that success in evolution has as much to do with maintaining the relative cohesion of parental 'instructions' for the development of the offspring as with the selection of individuals that are better able to reproduce in particular environments. It might not be fundamentally critical for organisms to be constantly and tightly adapted. There could very well be a much looser relationship between organisms and their supposed niches than we would suppose if natural selection were the only mechanism in town.

You will notice, Charles, that I've put 'message' and 'instructions' in inverted commas. This is because, strictly speaking, there is no blueprint or recipe in the DNA that can 'instruct' an organism how

to develop. This is an important aspect of biology that I will need to
return to once I've explained development within individuals.

Five thousand, but who's counting?

Let's take a pertinent example. There are more than five thousand
species of *Drosophila*. This means that there are five thousand
different ways of making an organism classified in the genus
Drosophila. We do not know whether all of these biologically incompatible types represents five thousand different modes of adaptation to
five thousand different environmental niches, forged solely by natural
selection. But we do know that the genetic 'instructions' giving rise to
all individuals in any one of the five thousand species are compatible
and that they involve hundreds of different types of TRAM systems.

I am proposing that each of these species is successful as a way of
life owing to the inherent ability to keep genetic 'instructions' on
how to develop and behave the same between newly coexisting
parental genomes during a period of evolutionary change in a sexually reproducing species. Accordingly, one criterion for success
would be the production of a multicellular organism capable of
reproducing with other similarly produced organisms, without the
need for every new variant to be constantly and selectively checked
by the environment. Naturally, it would be foolish to suggest that no
checking at all takes place. But it would be equally foolish, given
what we know about genomic turnover, redundancy and modularity
in biological processes, to ignore the fact that evolution can produce
successful individuals, judged as much by parental similarities in the
inherited genetic 'instructions' as by the forced contingencies of
survival in the local environment.

How might our five thousand *Drosophila* species have turned out
if molecular coevolution had taken different courses? If the turnover
events among redundant genetic elements had been slightly different, there might be a different set of five thousand species of
Drosophila from the set we have today. Each of these five thousand
virtual species might have survived just as successfully as the real
species, partly because sex and homogenization sustained genetic
compatibility during individual development.

I hope you appreciate that it is because of all these arguments that

I cannot give you a direct answer to your question about species formation as such. Biology's central task, in my view, is to sustain genetic compatibility during the natural life cycles of sexual organisms. Hence, within each evolving sexual population undergoing molecular drive and selection, there is a tendency to draw all individuals into sharing genetic systems that are successful in producing new individuals as a result of their similarity. If this also leads to the gradual build-up of differences between the genetic systems of separate populations, the ensuing biological incompatibilities might lead to new species. Species need to be viewed as the accidental by-products of separate paths of homogenization and cohesion in sexual populations, as well as the consequences of separate adaptations driven by the ecological criteria that govern selection.

The first sexual act

So far, I've been describing the consequences of sex for keeping everyone in line in the production of new progeny at each turn of the life cycle. This doesn't explain why the sexual alternation between diploid bodies (with paired chromosomes) and haploid sperm and eggs (with single, unpaired chromosomes) became established in the first place. Indeed, a satisfactory explanation of sex is still not forthcoming, despite all the thinking that has gone into it by experts such as John Maynard Smith. The difficulty is that it is generally assumed that having sex is not very clever for females if their sole aim in life is to leave as many of their genes behind as possible. Each mother would be better off with asexual cloning methods, producing only daughters identical to herself, which would guarantee that all her genes are passed on, for better or for worse. The problem with sex with males is that a female's genetic contribution to the offspring is reduced by half. This female-oriented argument only holds true if the male contributes no nutrients, through the sperm, to the egg, as is the case in humans. In which case males are expendable and the female is better off genetically going it alone.

To solve this problem it is assumed that there must be other advantages to sex, such as the long-term benefit of generating novel combinations of genes through the random shuffling of chromosomes by meiosis and mating, as I described in my very first letter.

This is as if sex is an insurance policy against any future adverse conditions. If things get bad, some of the genetic variation might turn out to be useful for future reproductive success.

I've never liked either the starting problem or its solution. The problem arises only if we assume that females have been manipulated to leave behind as many of their selfish genes as possible. You must recognize by now, Charles, that the evolution of genes involves far more than natural selection, so I will not labour this point. I'm also not comfortable with 'solutions' that rely on future events. Understanding life with hindsight is all too easy. As with the idea of adaptations being 'improbable perfections', hindsight can lead us to a misappreciation of what the real 'problems' were in the first place.

What we need to explain for the origin of sex is why two sets of chromosomes came together into the same cell and how and why they learned the trick of separating again. Neither of these events was necessarily difficult to achieve. Two chromosomal sets might have come together because a cell, originally with one set, failed to divide properly. Cell division takes place by a process called mitosis. Two sets of chromosomes would have the opportunity to reside in one nucleus if cell division were incomplete. Alternatively, two separated cells might have joined up together, using many of the existing proteins on their surfaces to join and stick together. Indeed, we know that many such activities, including the ability of cells to signal and communicate with each other, have evolved from cell-surface proteins currently found on single cells of bacteria. Needless to say, many signalling systems have the attributes of TRAM genetics.

The eventual separation of the two sets of chromosomes in the process of meiosis, to return the cells to the state of having one set of chromosomes (that is, forming the male and female sex cells), might not have been too difficult to evolve given that meiotic division is probably nothing more than a sort of faulty mitosis. In mechanical terms, the evolutionary switch from an asexual to a sexual life cycle might not have been too difficult to achieve.

The beauty of having haploid cells failing to separate completely after mitosis, coupled to their having two sets of chromosomes in the same nucleus, is twofold. First, we have the beginnings of multi-cellular organization; and second, those non-Mendelian mechanisms of turnover that can now get to work between the two sets of chromosomes. In other words, the transfer of genetic information can occur between homologous chromosomes, leading to elevated levels of homogenization between them. Homogenization, aided and abetted by the constant shuffling of chromosomes by sex, ensures that successful multicellular development of new offspring from the fertilised egg proceeds on the basis of similar parental 'instructions'. Seen in this light, the origins and persistence of sex have to do as much with the reduction of genetic separateness as with the generation of genetic diversity. Sex keeps everyone in line. The ability of organisms to develop into complex multicellular states and engage in sex (the never-ending cycles of haploid–diploid alternations in organisms above bacteria) probably evolved at one and the same time. The former might not have been possible without the latter.

Charles, before we leave this issue, I wonder if you could provide a suitable word to replace 'message' or 'instructions'. As I said, I use them in a very loose sense. The genetic material does not contain a set of instructions or blueprint for development. There is no 'start here' and 'finish there' program. Development is an unfolding of genetic interactions, starting with signals coming from the mother herself. These signals are part of the developmental unfolding of the mother that began with signals from her mother. And so on, back to the origin of life, if we want to be imaginative. The signals from the father occur at a slightly later time in the development of each individual, but the principle is the same.

So, what can I call this process of developmental unfolding that is specific to each species? The patterns of genetic interactions in a given species depend on the evolution of regulatory circuits that are TRAM systems. These patterns are therefore ultimately traceable back through evolutionary time, but they do not form any sort of recognizable 'instructions' in the DNA. Enrico Coen, in his book, *The Art of Genes*, likens the development of each individual to the creative act of painting the same picture over and over again. It is a result of a continuous interaction between the artist, the surroundings and the unfolding picture.

Naturally, I shall come back to these issues very shortly, but if you could help me out with a missing term, I would be grateful. What word would best describe a species-specific pattern of unfolding as one gene turns on or turns off another gene and so on, throughout development and through the generations? Maybe, Charles, you have some suitable nineteenth-century word that has long been forgotten.

In anticipation

Gabriel

Westminster Abbey

My dear Dover

I knew you would put me on the spot one way or another. Is this some form of test? I was always rather a plain-speaking man, as my writings and letters will testify, so I can't give you a fancy list of alternatives for the word you are looking for. But I can understand your dilemma.

How about 'mandate', as in 'the genetic material mandates the fertilized egg to develop in a direction specific to a given species'? 'Advice'? 'Guidance'? 'Unfolding'? (You've been using that word yourself.) I can only half help you because I only half understand the problem – you haven't yet explained how development proceeds. TRAM? Hox? Regulation? I'm still waiting for you to pull it all together now that I've been brought up to scratch. Please don't misunderstand me. I'm very grateful for your initiative in writing to me. I've certainly had my eyes opened to the nature of genes and their antics. I appreciate how you try to bring molecular knowledge into our age-old problems of adaptations, species formation and now, in your last letter, the origin and consequences of sex itself.

From what you say, it looks as if sex does have a dual effect of diversification and homogenization. There is no need, as you say, to pit one against the other because they represent two different aspects of the genetic material. In sexually reproducing species with alternating haploid–diploid life cycles, where the diploid state needs to develop from a single cell into a multicellular organism, the genetic uniformity of contributing parental genes is vital.

Clearly, the homogenizing mechanisms operating between chromosomes during meiosis will help to ensure that this is the case. The random mixing of parental chromosomes by sex will also ensure that, on average, there will be genetic uniformity among the new generation of diploid organisms.

The conclusion you draw from this, that the precise genetic system of guidance is not too important as long as there is uniformity in the two incoming guidance systems, is very interesting. I'm going to have to put some thought into that. If correct, your idea reduces the need to suppose that every important detail of species lifestyle is shaped by the forces that govern natural selection. It's going to take me some time to adjust to this. I am beginning to understand how the idea relates to what you said, all too briefly, about exaptations and adoptations. I hope we can discuss this further.

Ever yours affectionately

Charles

HOX! HOX! HOX!

Dear Charles

To begin to understand the mystery of evolution, we finally enter the wonderful world of *Hox* and other significant developmental genes. I can use these genes to discuss all the issues of redundancy, turnover, modularity and molecular coevolution.

Before I enter into this story, I'm ashamed to say that all my examples will be restricted to animals. But the spectacularly diverse world of plants could have supplied me with as instructive a range of examples as the world of animals. The molecular dissection of plant genomes and plant development has revealed that plant evolution is as much a consequence of genomic turnover among genetic units coupled to natural selection as is the evolution of animals. The two major kingdoms are no different from each other in the evolutionary processes that have shaped their separate biologies.

I'm happy to be able to say that my guilt over the omission of plants can be assuaged by the recent publication of one of the most perceptive books of the new developmental genetics, which includes many botanical examples. The book is called *The Art of Genes* and is written by Enrico Coen, one of the world's pioneers in unravelling the genetics of plant development. The title reflects Coen's delightful and imaginative connection between the artistic act of painting and the genetic act of development. I hope you find the time, Charles, to read Coen's book, which I consider the natural complement (and I hope he doesn't mind my saying so) to these letters.

'There is, philosophically speaking, only one animal'

This was the far-reaching conclusion of one of France's leading biologists, Etienne Geoffroy Saint-Hilaire, in the early part of the nineteenth century. Despite the ridicule heaped on him by that other stalwart of French biology, Georges Cuvier, Geoffroy has been more than vindicated by the late-twentieth-century discovery that the genes involved with animal development predate the origin of multicellular animals, perhaps one billion years ago. The multicellular animals include everything that the proverbial person on the Clapham omnibus would recognize as an animal – mammals, fish, insects, worms, crustacea, snails, nematodes, centipedes, leeches, sea-urchins – and many others that our bus user might not know were animals, such as corals, sea anemones, hydra and sponges.

From his anatomical studies of the two main animal divisions of vertebrates (with internal backbones and skeletons) and invertebrates (with no backbones but with external skeletons), Geoffroy was convinced that they were one and the same thing. Their different appearance was due to specific distortions in the processes of growth and development that made one look like an inside-out version of the other. As invertebrates predate vertebrates, this would mean that vertebrates are an inside-out version of invertebrates. The guiding concept that led Geoffroy to such a unity of types was his 'principle of connections' between basic body parts. These connections are capable of twisting and turning in many different ways, leading to bizarre configurations that look very unlike each other. Yet, unpick the developmental distortions of the connections in any given animal and the basic proto-animal will reveal itself.

That all animals are basically the same would be acceptable to any infant happy to draw a tube with a hole at one end for food and a hole at the other for waste. A slightly more advanced child would draw a head with eyes at the food end and some legs and perhaps wings sticking out in pairs from the sides of the tube. As far as the child is concerned, this cartoon would be as representative of what she saw in the mirror as of what she might see crawling around the bedroom. Apart from symmetrical or irregular forms such as sponges, corals, sea-urchins and star-fish, our clever child would have no difficulty in lumping insects and humans together. Neither did Geoffroy and neither do we.

Yet according to the history of science, Geoffroy's reputation suffered in the intervening 170 years because of Cuvier's insistence that the function of body parts was more important than their structure in the classification of animals. Cuvier's view has essentially held sway since then.

By the time you wrote *The Origin*, you could not avoid this debate. The 'illustrious Cuvier', as you wrote, was right to emphasize functional entities because they were, for you, the adaptive products of natural selection. But you also came to Geoffroy's aid by pointing out that the common bone structures of the human arm, the whale's flipper and the bird's wing signified a common origin. Their present-day differences were due to what you called 'Conditions of Existence' – that is, each was shaped by selection to perform a given function. Geoffroy's 'principle of connections' became subsumed into an all-embracing view of biology for ever providing functional solutions to problems imposed by an overpopulated, external world.

As I've discussed in an earlier letter, the evolution of all biological functions is perceived by many people as adaptations arising through natural selection. When I've finished telling the story of *Hox* and some other genes affecting the development of animals, I hope to be able to show you that structure and function are indeed intimately connected, but that both are also the result of continual turnover and coevolution among redundant and modular parts – that is, TRAM systems.

We could say that the only 'rationale' for many animals is that they work: they develop, eat, excrete and reproduce, in an environment that was not necessarily and primarily responsible for the structures and functions occupying its space. The process of producing something that works has to respect everything that has happened before on the branch of the evolutionary tree on which a novel life form may arise. We might be justified, however, in viewing every life form as much in terms of 'that's the way the cookie crumbled' as being the product of a relentless process of adapting in a harsh, competitive world.

Let me summarize this view, Charles, by asking you a question. Did all the legs on a centipede arise only through the inevitable provision, by natural selection, of novel solutions to external problems? Or could they also have arisen and spread by a molecular drive process in a small population of proto-centipedes, who, without too much fuss, got on with the business of living, reproducing and generally exploiting, with their newly acquired legs, some parts of their existing environment that were previously inaccessible? In short, did the centipede need to evolve all its legs?

This centipede example was not plucked completely out of the air. When I have finished the *Hox* story, I hope you will appreciate that I could have used many thousands of other successfully developing and reproducing animal life forms, which are a product as much of the internal flux of their genomes as of the external flux of the environment.

The first signs: shock, horror!

While you are thinking about these questions, Charles, let me give you my thoughts on the matter. To do this, I need to take myself back and you forward to 1894, which saw the publication of one of the landmark books in the history of genetics and evolution, *Materials for the Study of Variation* by the English zoologist William Bateson. Bateson coined the word 'genetics' and helped to establish a discipline that is set to revolutionize human medicine, and human society in general, hopefully for the better and not the worse.

Some of the most remarkable descriptions of natural variation in Bateson's book were of insects and crustacea in which a leg would be poking out of a fly's head instead of an antenna or an antenna in place of a crab's eye. Once one is over the shock of this, the real surprise comes with the realization that the leg in place of the antenna and the antenna in place of the eye were real legs and antennae exhibiting all the precise joints and segmental shapes of the particular species. They were not mixtures of legs and antennae or antennae and eyes: it was one or the other complete appendage. Presumably, some simple mutation has managed to undo hundreds of millions of years of evolution in a single instance and transformed the complete structure of the usual appendage into the complete structure of another appendage, normally found on a different segment of the body. Such transformations reveal the modular nature of appendages that can develop out of position without disturbing the rest of the body.

Bateson called this type of variation 'homeosis', from which we derive the terms 'homeobox', 'homeodomain' and *Hox* genes, among others, representing the fruits of labour of geneticists working in a wide range of animals and plants. The pace of discovery is relentless: we are on the threshold of understanding some of the deepest secrets of why biology evolved the way it did. But poor old Bateson knew nothing of this, despite his prophetic comparison of the discovery of homeosis in animals (and of Goethe's similar findings in plants) with the advent of the prism for understanding the seven colours of white light.

It has been persistently argued since Bateson's time that such 'gross' mutants, for that is how they were viewed, could not be the real stuff of evolution. Evolution required very small, almost imperceptible changes, involving thousands of separate genes and accumulating over vast periods of time by natural selection. You made the same point to me in one of your letters, where you pitched 'gradualism' against 'punctualism'.

I would love to know what your reaction might have been had you observed such homeotic transformations. Would you have realized that they were trying to tell us something very important about how organisms built themselves from modular components, under the control of relatively small numbers of genes? Would you have adjusted your view of natural selection as being capable of working only on small differences one bit at a time, so that if we looked hard enough we would find all the transition stages of antenna/leg mixtures as antennae turned into legs? And would you still have believed that each mixture had a beneficial effect on the relative reproductive success of its bearers?

Legs and antennae come in all shapes and sizes. We just have to think about the arthropods (insects, crustaceans, centipedes, butterflies, grasshoppers and so on) to get a glimpse of the level of diversity. Maybe the evolutionary transformation of one type of leg to another, or even the whole appearance or disappearance of an appendage in a given species, is the result of homeotic processes involving a few 'master' genes rather than hundreds of genes each with a small effect.

The real issue here might not be the eternal debate between a few genes and many genes, or even between large genetic effects and small genetic effects. The central problem revolves around the adaptive value or not of intermediate stages. For natural selection to work, intermediate steps have to confer greater reproductive success. Hence, it is legitimate to ask, in a model of natural selection acting alone, what use is half a wing? To which we are forced to answer: whatever use gave the bearers of half wings a reproductive advantage. You called these intermediate features 'preadaptations', which neatly sidesteps the relevance of the question, 'what use is half a wing?' What 'preadaptation' stresses is the necessity of improved adaptive values at all transition stages of natural selection, even if such values remain mysterious.

Bateson's homeotic mutants have, belatedly but finally, opened our eyes to alternative criteria by which evolutionary success of biological innovations can be judged. This bold statement is open to misinterpretation, so I will plunge straight into the genetics of homeosis and its role in the successful diversification of form and function.

Beyond head, middle and tail

Our young infant drew the head, middle and tail of all animals: the proto-animal. When he starts to explore the garden, he will observe that butterflies have two pairs of wings, and bluebottles have only one pair of wings at the front, along with a pair of very small wing-like projections called halteres just behind. Eventually, if he attends the right university course, he could be shown 200 million-year-old, four-winged fossils, indicating that the four-winged butterfly is the primitive state, and that the hind pair of wings was subsequently reduced to halteres, used for balancing during flight.

Imagine the surprise of our student, now a low-paid working geneticist, on finding halteres transformed into wings in some individuals of our favourite fly, *D. melanogaster*. Sometimes the transformation involves the top (anterior) part of the haltere, and sometimes it involves the bottom (posterior) part; and sometimes mutations are uncovered that give rise to four-winged flies! That's another 200 million years of evolution undone in the blinking of an eye.

We can now give a name to our researcher and call him Ed Lewis, and follow his progress from his first genetic studies of homeosis in the 1960s through to the award of a Nobel Prize in 1995. The intervening three decades witnessed a long and persistent study of the 'homeotic' genes, most of the results of which were ignored by the arbiters of evolutionary taste. Lewis's discovery of the first *Hox* genes, however, was to set in motion not just a major reconsideration of the genetics of development, but also a considerable re-evaluation of our perceptions of the evolutionary forces behind the successful establishment of biological novelties.

Naturally, Lewis could play only one role in the emergence of this alternative biology. There are several other key biologists involved in the unfolding story, but I do not believe it necessary to name at each and every turn. Instead, Charles, I'm sending you, under separate cover, many of the relevant papers and books that have been written. I hope the custodians of the Abbey don't fret too much about their eventual delivery.

Masters and slaves

The ease of transformation of a modular leg for a modular antenna, or a modular wing for a modular haltere, shows that the developing embryo has to make some very important choices at some crucial steps in development. If it makes the wrong choice, a perfectly natural and familiar appendage appears in an unnatural place. We can think about the underlying genetics of this in three ways.

One way is to assume that these choices involve a gene for a leg, or a gene for an antenna, or a gene for a wing, and so on. With this way of thinking we could progressively account for every aspect of an individual's appearance or behaviour by a gene whose sole purpose is 'for' a particular slice of that individual. This route quickly leads us to the belief in a gene for sexuality, a gene for language and a gene for aggression, about which I've complained in the past and no doubt will complain again in the future.

A second way to think about the homeotic observations is to suppose that there is a 'master' gene controlling a set of 'slave' genes that collectively make a leg, or an antenna, or a wing and so on. By 'controlling' I mean turning other genes on or off by producing regulatory proteins that bind to the respective modular binding sites in the promoters of 'slave' genes. Homeotic transformations could then be seen as the result of simple mutational switches in the function of the master gene, as they control alternative sets of 'slave' genes.

Whatever the precise details might be, we are still assuming that there is a particular 'master' gene with its particular set of 'slave' genes that have evolved for a specific purpose – for the production of a wing or a leg and so on. All we have done is to widen the initial abstraction of a specific gene for a specific body part to one of a specific master/slave package for a specific body part. This expansion is more realistic than the first possibility, but it turns out to be wrong.

Promiscuous genes

There is a third way of thinking about homeotic transformations that dispenses with the words 'specific' and 'particular'. There are indeed genes that have been called 'master' genes, in that they regulate what I have called 'slave' genes, or what are frequently called 'target' genes. But neither the 'master' genes nor the 'slave' genes are uniquely restricted to performing one function only, such as leg development or wing development. Most of the genes involved with the leg or wing can be shown to participate in many different functions both within a given species and between different species. Recent evidence has emerged that genes are unexpectedly versatile in their contributions to unrelated developmental, metabolic and behavioural processes. Now, given that a gene makes its contribution to a function or structure through its molecular interactions with other genes and their protein products, the versatility of the gene points to a promiscuity of interactions. A gene can bed down with multiple partners and it is the combination of partners, in any given cell at any given time during development, that gives rise to a particular part of the individual.

With this third scenario, no particular gene, or even complex of genes, exists uniquely for a given biological function. But a given species-specific structure can emerge at the right time and in the right place when a group of multifunctional genes is active or inactive at a given locality and point in time in the appropriate combination. The coincidence of gene activities at one stage of development is contingent on earlier coincidences of gene activities often involving some of the same genes. If this sounds confusing, I shall give you some real examples shortly.

How did a particular group of multifunctional genes combine to produce a wing or leg? If evolution is considered to be the result of selection alone, acting directly on this or that selfish gene to promote adaptational functions, then we have a problem on our hands. Which particular function of any one of our multifunctional genes does selection favour at any one time? The only way out of this impasse is to consider selection acting on phenotypes, or parts of phenotypes, in exactly the way you proposed. Phenotypes are the shifting allegiances of interactive genes. The individual product of such local interactions is central to our understanding of the incessant diversification of form.

If we take this starting assumption as realistic, to understand how and why combinations of genes give rise to certain structures we need to consider where all the flexibility and promiscuity comes from. As you might guess, Charles, the answer lies in the evolutionary dynamics of TRAM systems. Well, let's take this one step at a time. How do we know that a wing or leg in *Drosophila* is the result of a particular combination of versatile genes coming together at a given time and place during development?

Serial transformations

An adult fruitfly develops from a segmented larva, the last stage of which has fourteen segments. The relationships between the larval segments and the adult segments are shown in the last two diagrams. I have labelled the segments according to their adult function. For example, the three segments making up the thorax are called T_1, T_2 and T_3. Each of these segments carries a pair of legs, while T_2 also carries a pair of wings and T_3 also carries a pair of halteres. Segments A_1 to A_8 represent the abdominal segments. The head is also made up of segments, but I do not intend to discuss these further.

Ed Lewis found that the same genes that caused homeotic transformations in the adult (for example, a pair of wings on T_3 in place of halteres) also caused segment transformations in the larva.

Three particular genes, which turned out to be among the first three *Hox* genes discovered, could be deleted from the chromosome and the resultant segments of the embryo examined. These genes are called *Ultrabithorax* (*Ubx*), *abd-A* and *abd-B*. If all three genes are deleted together, which is possible because they lie very near to each other on the same chromosome, the resultant larva has segment structures T1, (T2) × 10. In other words, ten of the segments have the structure of T2. If *Ubx* is added back, the larva has segment structures T1, T2, T3, (A1) × 8. If *abd-A* is added back as well, we have T1, T2, T3, A1, A2, A3, (A4) × 5.

In the absence of all three *Hox* genes, all the segments except T1 would turn into T2, and if the larva could overcome this little problem and develop into an adult, we would have a fly with ten pairs of wings and eleven pairs of legs – something resembling Manchester United Football Team, to which I shall return later! In the absence of only the *Ubx* gene, we get our four-winged fly. In the absence of *abd-B*, several of the most posterior segments resemble the fourth abdominal segment. And finally, in the absence of both *abd-A* and *abd-B*, all the abdominal segments look like the very first one. How can we explain these serial transformations?

Putting it all together by modules

Lewis proposed that each gene is turned on in a specific territory across several segments and, importantly, that the three territories overlap like steps of a terrace.

For example, segment T3, which normally carries the pair of halteres, expresses, according to the model, only the *Ubx* protein out of our three *Hox* genes. So we can conclude that the absence of the *Ubx* protein (in the mutant fly) allows a pair of wings to develop in place of the halteres. This is similar to the permanent absence of the *Ubx* protein in the adjacent segment T2, which consequently always develops into a pair of wings. From this we can deduce that one of the normal purposes of the *Ubx* protein is to suppress the formation of wings in segment T3.

All in all, eight *Hox* genes have been found in *Drosophila* that are expressed in overlapping domains along the head–tail axis of a developing embryo. One of these *Hox* genes (called *Antennapedia*)

Drosophila larvae   Hox genes

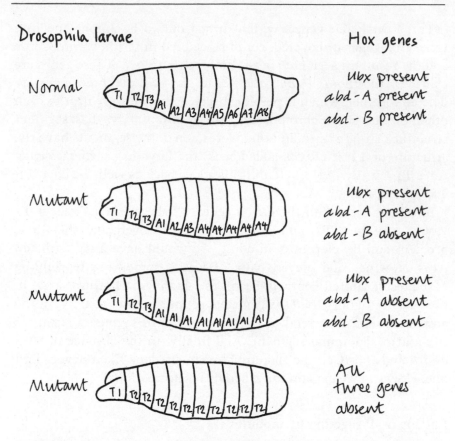

Normal   Ubx present
                                                     abd-A present
                                                     abd-B present

Mutant   Ubx present
                                                     abd-A present
                                                     abd-B absent

Mutant   Ubx present
                                                     abd-A absent
                                                     abd-B absent

Mutant   All
                                                     three genes
                                                     absent

Segmented larvae of the insect *Drosophila melanogaster* *The normal (non-mutant) larva form shows three segments (T1, T2 and T3) that will develop into the segments of the adult thorax (with antennae, wings, legs and halteres; see next diagram). The rest of the segments (A1 to A8) give rise to the adult abdominal segments. The three mutant larvae show the state of segments when one or other of the three Hox genes (Ubx, abd-A; abd-B) are removed, singly or together. For example, when all three Hox genes are missing, all segments except the first are T2 in structure. If this larva could survive to adulthood, it would give rise to a fly with eleven pairs of legs and ten pairs of wings!*

turns out to be responsible for promoting the formation of legs in their usual places. Hence, should the gene be expressed in the part of the head where an antenna is normally produced, then an antenna to leg homeotic transformation takes place.

Antonio Garcia-Bellido from Madrid, a leading investigator of genetic interactions during development, proposed to call these powerful *Hox* genes 'selector' genes because they can select out for activation or repression many other genes whose business is to

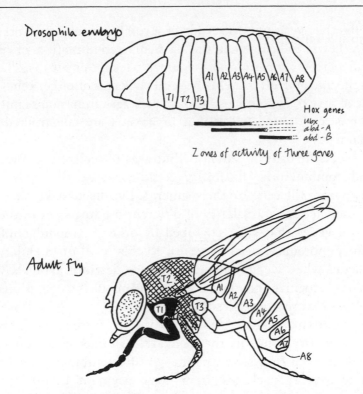

Drosophila embryo

A1 A2 A3 A4 A5 A6 A7 A8

T1 T2 T3

Hox genes
Ubx
abd - A
abd - B

Zones of activity of three genes

Adult fly

This diagram shows the relationship between the larval segments and the parts of the adult fly. T1, T2 and T3 represent the three segments of the thorax and A1 to A8 are the abdominal segments. The bars represent the extent to which each of the three Hox genes (Ubx, abd-A and abd-B) is turned on in the larva.

produce a leg or an antenna. The genes under the direct control of the 'selector' genes were called 'realizator' genes. These terms are interchangeable with the terms 'master' genes and 'target' genes. What they are called is irrelevant to the central idea that batteries of 'target' genes are under the negative or positive control of a few 'master' genes. In other words, the particular pattern of 'on' and 'off' switching in the set of target genes determines the production of, for example, a leg or antenna. The 'target' genes come as packages of genes whose interactions ultimately lead to the formation of an adult appendage, and the particular interactions in any given appendage depend on the 'on/off' switching patterns produced by the 'master' genes.

Finally, the overlapping domains of several 'master' (selector) genes seen by Lewis indicate that the particular combination of master genes in a given segment (a sort of super 'package of packages') determines the 'on' and 'off' states of several packages of target genes. The genetic ability to switch master/slave packages from one segment to another is a clear indication that organisms are constructed from modular parts.

The special insight of Garcia-Bellido was to realize that modularity and combinatorial flexibility could exist at all levels of an organism from the genes to the segments. I've discussed in an earlier correspondence the modularity of different binding sites in the promoters of genes that are recognized by different modular folds of regulatory proteins. I need to return to this level of modularity in more detail when we start asking serious questions about how the *Hox* genes recognize their target genes and about the operations that control the *Hox* master genes themselves.

The important level of modularity in our current discussion of segment identity concerns the 'package' of target genes under the control of a given *Hox* gene. It is the specific combination and interaction of several 'packages' in a given segment that ultimately determines the structure of the segment. Garcia-Bellido has proposed that ultimately the individual cell is the minimum autonomous unit that emerges from a given combination of target gene packages. Each package is involved with a particular developmental process and the package, along with its process, is a module of information that can be combined with other packages and processes in specific cells.

Some of these genetic packages induce developmental processes through which cells can signal to each other. This ensures that they respond, as a cohesive collection of cells, to the presence of other genetic packages. By such means phenotypes develop from cell to tissue to organ. And throughout this process individual genes of this or that package partake in many different developmental pathways. It is the particular combination of active genes that determines the direction of a particular pathway, and not pathway-specific genes.

Given these complexities, it is difficult to define the precise boundaries of a genetic package. It could be an individual cell, part of a segment (as I shall show below) or a whole segment, with its corresponding adult appendage. For now, what is important to realize is that genetic packages can be called up or dismissed as autonomous modules in different cells during normal or abnormal development. For example, the genetic package under the control of *Antennapedia* can be turned on in the cells usually producing the antenna, leading to the production of a leg. This shows that the *Antennapedia*-controlled leg package is already present in the cells normally producing an antenna. But it does not normally produce a leg because the *Antennapedia* master *Hox* gene is not normally 'on' in such cells.

Before and after *Hox*

The *Hox* genes have to be understood in the context of what comes before *Hox* and what comes after *Hox* in the hierarchy of events that takes place as an egg matures into an adult.

Lewis's and Garcia-Bellido's interpretations of the control that *Hox* genes exert over other genes have been fully borne out at the molecular level. As I have explained previously, *Hox* genes produce proteins containing the sixty amino acid module called the 'homeo-domain' that is capable of binding to specific short DNA modules in the promoters of the target genes. How does any given subset of these promiscuous, multifunctional target genes come under the control of a *Hox* gene?

Let's look at this question with respect to *Ubx*, whose role is to prevent a second pair of wings developing on the third thoracic segments (T3) in a fly. Instead a pair of halteres is made. In the absence of *Ubx*, a four-winged fly emerges. The wing/haltere developmental choice is relatively simple to study because only one *Hox* gene, *Ubx* is expressed in the larval cells that will develop into a haltere.

A mess, but it works

The next stage gets a little complicated. Charles, I have thrown into the wastebasket several pages of detailed description of what is known of the differences in genetic interactions between the package of target genes producing a wing and the package producing a haltere. The genes involve at least eleven of the best-known genes in *Drosophila* development, many of which have other functions elsewhere in a developing fly. For example, they include genes that

produce regulatory proteins, or proteins bound within cell mem-
branes, or the important signals by which cells communicate one
with another. I've decided against presenting all these details
because they are available in some of the books and papers that I've
forwarded to you. The overall message is that the *Ubx*-controlled
switch of haltere to wing involves *Ubx* interference during several
stages. Furthermore, the eleven key *Ubx* target genes known to date
are probably just the tip of the iceberg. According to calculations by
some experts on *Drosophila* development, about half the genes
known to be involved in building a wing are, directly or indirectly,
suppressed by *Ubx* in the haltere, making about thirty *Ubx* target
genes. If we also take into account many other differences between
the T2 (wing) and T3 (haltere), including the peripheral nervous
system, muscles and trachea, as well as *Ubx* involvement in the
central nervous system, there may be up to a thousand target genes
for any given *Hox* gene.

As I have emphasized before, the *Hox* genes, like many of their
target genes, are highly promiscuous. They produce proteins with
homeodomain modules that can recognize homeodomain binding
sites in the promoters of a wide range of genes. Indeed, the promoters
of many genes are compound structures in which one section is
involved in expressing a gene in one place and another section in
another place.

All in all, what we are seeing is extensive sharing of genes between
different tissues and the development of each tissue as a result of
interactions between packages of shared genes that are specific to
each tissue. Any given set of interactions is influenced by the particular combination of *Hox* master genes being expressed in a given
tissue. Finally, *Hox* genes, like many other developmental genes,
are multifunctional in that they can be involved with unrelated
developmental processes.

Before moving on to the central question of how any of this arose
during evolution, I want to outline the genetic circuitry that controls
which *Hox* genes are on or off during development.

From 0 to 14 in four hours

The egg is a single cell with a single nucleus containing paired chromosomes. It is formed from the fusion of the male and female nuclei, which contain unpaired chromosomes. Four hours later, the proteins of several genes, called the 'segment polarity' genes, accumulate into fourteen transverse stripes along the length of the embryo, looking something like the black-and-white keys of a piano. These are the future segments of the larva. *Hox* genes in turn are regulated with respect to some of these boundaries, which control the combination of target gene packages in each of the segments.

If we go back in time to an earlier stage, the embryo is divided up into seven stripes depending on the distribution of proteins among the 'pair-rule' genes. And prior to the seven stripes stage, several 'gap' genes are involved in the internal subdivision of the original one-cell egg into top and bottom. Indeed, the 'gap' genes themselves are controlled by the proteins of several genes of the mother herself that have entered the egg.

All this appears to be perfectly sensible and logical. There is a hierarchy of gene activity with one layer controlling the expression of another as the egg is subdivided into fourteen domains, each of which develops into a *Hox*-controlled segment. Naturally, there is some truth in this, but the network of interactions between all the relevant genes is not at all explicable in terms of a simple hierarchy of 'on' and 'off' switches, from the maternal genes to the *Hox* genes.

Getting lost in the net

I've drawn a simplified version of some of the interactions to show just how complex the situation actually is. There are several interesting features. First, as I've explained, many regulatory genes have multiple functions and each function can be of two types: a given gene can turn another gene either 'on' or 'off'. If you look at the 'gap' gene *hunchback* (*hb*) in the drawing, you will see that it can turn 'on' *even-skipped* (*eve*) and *runt*, and turn 'off' *hairy*.

Unfortunately, this diagram can illustrate only a two-phase system of 'on' or 'off'. What it cannot show is the quantitative control of one gene over another – like a variable light switch that controls how bright you want the bulb to glow.

The second feature of the diagram is that a given gene can be turned 'on' or 'off' by a number of other genes, all of which appear to be doing the same thing. Look at the 'pair-rule' gene *hairy*. Four other genes, *runt, hunchback* (*hb*), *Kruppel* (*Kr*) and *knirps* (*kni*) turn this gene off while *even-skipped* (*eve*) turns it on. This is one example of the redundancy in the system: that is, several genes apparently doing the same job. Another important level of redundancy involves the multiplicity of binding sites in the promoters of all these genes to which regulatory proteins target themselves.

Finally, although my diagram might look complex, if I put in all the lines of known interaction between the genes, the whole thing really would look a mess. For example, *bcd* is known to regulate *Kruppel* (*Kr*), *knirps* (*kni*), *even-skipped* (*eve*) and several other genes, in addition to *hunchback* (*hb*) and *giant* (*gt*) as shown in the diagram. Such extensive sharing is the result of the promoters of unrelated genes carrying *bicoid*-binding sites. The same is true of other regulatory genes that have a multitude of targets. This sharing in turn indicates that some mechanism of turnover, such as transposition or gene conversion, is involved in the spread of modular binding sites among promoters.

Extending the net

For completion we need to connect the genetic processes in my diagram with the *Hox* genes. At the bottom of the diagram I've just shown the regulatory connections to two of the *Hox* genes. There are eight *Hox* genes altogether, so you can imagine how complex the net would look if I connected all eight to their respective 'control' genes.

As you might expect by now, biology never takes the easy route. Several of the genes, particularly the 'gap' genes and 'pair-rule' genes, directly affect the 'on' and 'off' switching of the *Hox* genes. And interactions among the *Hox* genes themselves are part of the overall regulation of *Hox* expression. For example, the *Hox* gene *Antennapedia* has a compound promoter consisting of two regions. The first region is stimulated by *Kruppel* (a 'gap' gene) and repressed by *Ubx* (a *Hox* gene), while the second region is turned 'on' by *hunchback* (a 'gap' gene) and *fushi-tarazu* (a 'pair-rule' gene) and turned 'off' by *oskar* (a maternal gene). Is there any sense to this?

So you see, Charles, there isn't a simple graded hierarchy of gene expression with the first level controlling the second level and so on.

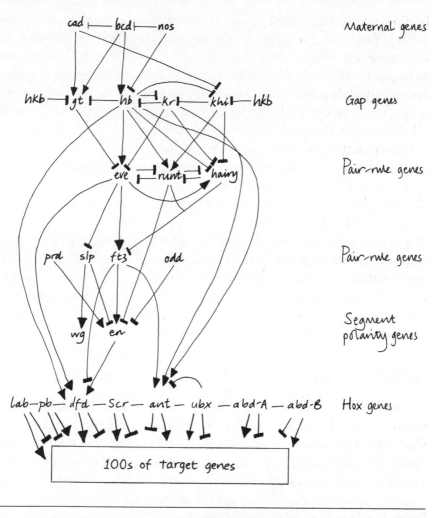

cad — bcd — nos   Maternal genes

hkb — gt — hb — kr — khi — hkb   Gap genes

eve — runt — hairy   Pair-rule genes

prd   slp   ft3   odd   Pair-rule genes

wg   en   Segment polarity genes

lab—pb— dfd — Scr — ant — ubx — abd-A — abd-B   Hox genes

100s of target genes

———▶ Gene turned on   ———▌ Gene turned off

Regulatory gene interactions in early developmental stages of *Drosophila melanogaster* *Genes are classified into maternal genes, gap genes, pair-rule genes, segment polarity genes, Hox genes and target genes according to their role during development. I can only depict part of the complex web of interactions in which one gene can turn 'on' or 'off' a variety of other genes. The set of interactions, specific to a given species, feeds through to the regulation of eight* Hox *genes in this species. Only the regulation of two of these genes is shown. The* Hox *genes in turn regulate hundreds of target genes that ultimately go towards the development of cell and tissue structures, such as wings and legs. Note that a given regulatory gene can affect a number of other genes, each of which produces a regulatory protein in its own right. The promiscuity of gene activities is due to the modular and redundant nature of promoters undergoing turnover, as seen in my earlier diagrams.*

Hox genes are controlled by genes drawn from several stages in the network of regulatory genes, as far back as the maternal genes and including genes from other systems entirely. Frankly, it's a mess, but it works in making the species *D. melanogaster*. I am arguing that the mess could have been otherwise, producing another viable and reproductively successful *Drosophila* species if the three processes of evolution (natural selection, neutral drift and molecular drive) had taken slightly different turns. Success can be measured by tolerance and flexibility between all purpose keys and multiple locks.

Hox is everywhere

What is fascinating is that the *Hox* genes are conserved across all animal groups, but some of the genes controlling *Hox* can be very different, as are some of the target genes controlled by *Hox*. The implications of this are that evolution of form and behaviour depends on the evolution of promoters and regulatory proteins. From what we know about some of these systems in other species of arthropods, there is no reason to believe that egg and embryo development follows exactly the same routes through the interactive mess as we see it in *D. melanogaster*.

For example, one of the all-important maternal genes called *bicoid*, at the very top of the hierarchy, has not been found in beetles and other arthropods outside of the dipteran fly. But many of the other genes are shared by animals as far apart as mammals and worms. The evolved differences between animals result from the specific combination of subsets of genes establishing networks of regulatory interactions as development unfolds. Such changing patterns of regulation ultimately imply a continual restructuring of gene promoters on an evolutionary timescale. Because all known promoters have the features of TRAM genetic systems, their evolution involves an interesting mix of molecular drive and natural selection. So we can expect comparisons of the same promoter between species to show that molecular coevolution has taken place between the DNA-binding modules of the regulatory proteins and the multiple binding sites in the promoters.

Looking for molecular coevolution in development

Has molecular coevolution occurred, for example, between *bicoid* and its binding sites in the *hunchback* promoter? Can we expect to find that within each species there is perfect compatibility between the *bicoid* homeodomain and its binding sequences in the *hunchback* promoter, but incompatibility between the *bicoid* homeodomain and binding sites taken from two different species?

Knowing in advance where to find evidence of molecular coevolution between the *bicoid* and *hunchback* is not easy. Remember that in our earlier discussion I was reluctant to say that molecular drive is a process of species formation. Given the huge variety of genetic systems with TRAM features, it is impossible to say which is responsible for the biological incompatibilities of two species. We are back to the ripples in the pool: which ripple of molecular drive is responsible for the inception of some critical aspect of species difference? Even in the case of molecular coevolution, we have no reason to expect this to show between pairs of the most closely related species.

Interestingly, one of the first examples of molecular coevolution among promoters of developmental genes involved the two genes *fushi-tarazu* (*ftz*) and *bicoid*. Both have the sixty amino acid homeodomain module, but at amino acid position number 50, the *ftz* protein has the amino acid glutamine and the *bicoid* protein has lysine. There is a corresponding difference in their respective binding sites in the promoters of several other genes. The *ftz* binding site has a pair of CC nucleotides at the start (CCATTA), whereas the *bicoid*-binding site starts with GG in the same positions. When the glutamine of *ftz* is replaced artificially by the lysine of *bicoid*, the *ftz* protein seeks out the GG-containing binding sites of *bicoid*. Clearly the glutamine-CC and the lysine-GG couplings are both perfectly functional, indicating molecular coevolution. Given the multiple binding sites shared by unrelated promoters, we can expect molecular drive to have spread new variant sites leaving selection to promote compensatory mutations in the homeodomains.

Another example

In the case of the *bicoid–hunchback* interaction, there are some
obvious upper and lower boundaries to the possibility of molecular
coevolution. We know that *bicoid* does not regulate *hunchback*
outside the species of the true flies, the higher diptera. So there is no
point looking for molecular coevolution between *D. melanogaster*
and a butterfly, grasshopper or beetle species. We also know that in
the genus *Drosophila* no amino acid differences have been observed
in the *bicoid* homeodomain region between *D. melanogaster* and *D.
pseudoobscura*, an evolutionary gap of some sixty million years.
Molecular coevolution might therefore require us to go further out to
genera that have been separated from *Drosophila* for over sixty
million years, yet still remain within the higher diptera.

Well, to cut a long story short this is what was done in the examination of the *bicoid–hunchback* interaction in the housefly *Musca
domestica*. This genus diverged from *Drosophila* about a hundred
million years ago.

Are there differences in the *bicoid* homeodomain between *D.
melanogaster* and *M. domestica*, and are there corresponding differences in the *bicoid*-binding sites in the *hunchback* promoter? The
answer is yes for both questions. There are five different amino acids
in the stretch of sixty amino acids that make up the homeodomain, a
relatively high number considering the sixty million years of homeodomain conservation in the genus *Drosophila*. At the promoter end
of things, there are ten *bicoid*-binding sites in the *M. domestica*
promoter, compared with six sites in the *D. melanogaster* promoter.
And the sequences, positions and orientations of the sites differ
between the species. The binding sites are not identical in sequence
in either species; however, there are, on average, some consistent
differences in binding sequence between the species.

Is there a causal link between the two divergent components?
Have the *bicoid* homeodomains diverged in response to an inherent
restructuring by turnover of the *hunchback* promoters? It is not easy
to establish causal links between given components in real organisms because of the high numbers of other genes involved with any
given interaction. Many such confounding factors can be removed by

conducting experiments in test-tubes, although these have the disadvantage that they do not reflect the true biological situation in all its gross complexity. Nevertheless, test-tube analysis of the relative binding affinities of *bicoid* homeodomains and *bicoid*-binding sites in *Drosophila* and *Musca* reveals that the *bicoid* homeodomain of *Musca* has a preference for the average binding-site sequences of its own promoter, rather than for those of *D. melanogaster*.

Is this preference a sign of molecular coevolution? Another hint comes from the subtle differences in the patterns of activity of the *Musca hunchback* when injected into mutant embryos of *D. melanogaster* that lack their own *hunchback* gene. This too could reflect the accumulation of differences in recognition between the *bicoid* homeodomain module of one species and the *bicoid*-binding sites in the *hunchback* promoter of another species.

Interestingly, the sequences surrounding the *bicoid*-binding sites in the promoters indicate that widespread slippage has occurred in this region in both species. Slippage is the non-Mendelian mechanism that produces lots of short repetitive sequences, sometimes called 'runny' DNA.

So what may have happened in the evolution of the *bicoid*–*hunchback* interaction follows on from the example of rDNA and *period* gene molecular coevolution that I've described earlier. The presence of runny DNA and fluctuations in the position, orientation and number of copies of *bicoid*-sites show all the hallmarks of slippage and unequal crossing-over. And the differences in sequences between the set of binding sites in *Musca* and the set in *Drosophila* indicate that there is some mechanism (maybe gene conversion) spreading mutations among the binding sites, which are specific to each species. There is quite a high level of divergence in the *bicoid* homeodomain between the two species and there are measured differences in binding affinities between the relevant components taken from the two different species. All this suggests that natural selection has responded to the spread of new, differently structured promoters by molecular drive. In other words, molecular coevolution between the homeodomain and the promoter might be taking hold.

I am not suggesting that molecularly driven changes among the redundant binding sites, followed by selection for compensatory mutations in the homeodomain, make all the difference between

D. melanogaster and *M. domestica*. There must be hundreds of accumulated genetic differences between these species, involving differences in the usage of other TRAM systems. Remember I told you about the *Odysseus* gene involved with male fertility and species formation in *Drosophila*. Its homeodomain has fifteen out of sixty amino-acid differences, even between closely related species. No fly, therefore, or population of flies woke up one morning to find itself turned into a *Musca* housefly because of the potential coevolutionary changes between *bicoid* homeodomain and *hunchback* promoter binding sites.

Kafka's hero Gregor may have woken up in the shape of an insect in the story *Metamorphosis*, but this cannot happen in real life. The evolutionary distances involved are just too large. Similarly, there can be no doubt that there are many other evolutionary steps between *Musca* and *Drosophila* that reflect turnover and selection in all of the relevant genetic systems. Let's take *bicoid* as an example. It regulates perhaps another ten genes apart from *hunchback*. We cannot fall into the trap of isolating the coevolutionary events between *bicoid* and *hunchback*, in ignorance of the knock-on effects on the other genes. Sorting out biology is a major headache.

The eyes have it

One of the important features of regulatory networks involved with development is the ability of a master gene, such as *Hox*, to switch from one package of target genes to another. I explained this briefly using the switch from antenna to leg packages by the *Hox* gene *Antennapedia* in *D. melanogaster*.

A dramatic example of the potential of cells to have some of their packages of target genes turned 'on' when they are usually 'off' came from the experimental shifting of the master gene, *Pax-6*, from one place in a developing fly to another. *Pax-6* is a homeodomain coding gene that controls a suite of genes leading to the formation of an eye. The *Pax-6* gene of mammals is interchangeable with the equivalent *Pax-6* gene of *Drosophila*. The experiment of getting *Pax-6* switched on in tissues not normally expressing *Pax-6* can be done using either the fly's own *Pax-6* gene or one taken from a mouse. Either way, the results are quite startling – eyes are produced on several unusual parts of the fly's body, such as legs and wings.

Insect eyes and mammalian eyes are constructed in different ways. The insect eye is a many-sided compound eye in which each 'eyelet' and its lens can be said to be equivalent to the single mammalian eye. So what should we be expecting from our gene-swopping experiments? We might guess that the *Pax-6* of *Drosophila* produces patches of compound eye wherever *Pax-6* is operating. But what of the *Pax* master gene from a mouse? Will it produce an insect or mouse eye when turned 'on' in different *Drosophila* tissues? The answer is that it produces a *Drosophila* compound eye on legs, wings, halteres and antennae. Moreover the antenna eyes generate the same electric impulses as the normal eyes showing that they are functional. Eyes cannot be induced in any tissue; the foreign eye master gene has to be active at the appropriate stage of insect development. Nevertheless, the mouse *Pax-6* can successfully interact with the estimated 2,500 target genes lying dormant in many different *Drosophilia* organs. *Pax-6* genes exist in such diverse species as amphibians, squid, worms and planariums, as well as mammals and insects. Some of these have also been shown capable of inducing extra eyes in *Drosophila*.

What these experiments tell us is that *Pax-6* really can be considered a master control gene that has been conserved over a huge distance of evolutionary time, probably predating the Cambrian epoch of 550 million years ago. But the 'package' of target genes is not the same between insects and mammals. There is no 'mouse' package in *Drosophila* waiting to be turned on by mouse *Pax-6*. This does not mean, however, that the mouse 'package' consists of an entirely different set of target genes from those in *Drosophila*. As I explained earlier, individual genes in a 'package' can be shared across widely divergent species. What distinguishes the mouse 'package' from the *Drosophila* 'package' is the combination of genes. The set of genetic interactions taking place in the mouse 'package', over space and time in a developing mouse, are different from the set of interactions taking place in the *Drosophila* 'package', even though many of the genes are the same. We can expect there to be some additional genes involved with a mouse 'package' that are unique to mouse, and likewise for the fruitfly, but this simply reflects the multifunctional nature of genes, as I explained above.

I'm sure that when the full glory of all the genes in all the modular 'packages' of development and behaviour are exposed, no large numbers of species-unique or package-unique genes will be found. Instead, differences in the modular construction of organisms will be the result of specific permutations of universally shared modules. For example, some of the genes that are known to interact with *Pax-6* in *Drosophila*, such as *twin of eyeless, sine oculis, eyes absent, dachshund, eyegone* and *teashirt*, are involved in the development of the sex gonads, legs and embryo segments. And there is no simple linear pathway for eye development starting with *Pax-6*. As with the genes influencing embryo development that I described previously, there is a complex network of interactions involving multifunctional genes regulating each others' activities via versatile, modular and compound promoters.

Each subsection of a compound promoter must contain the binding sites to the regulatory proteins of other genes which ensure that the gene in question is expressed in a given place during development. But this is not the end of the story because once a developmental gene is expressed in a new position it has to acquire the ability to interact with a different set of target genes. In other words, developmental genes have been co-opted, or recruited, to participate in other developmental processes, recruitments which could very well make the difference in the development of different species. But given the promiscuity of development genes, it is not possible to assign exclusive formative roles to many development genes.

The *Pax-6* story tells us that there has been just one origin and one evolutionary line of progression, from the earliest patches of light-sensitive cells to the variety of advanced eye-forms around us. This unavoidable conclusion, Charles, goes against a hundred years of insistence that the widely different structures and operations of eyes (eye cup, pinhole, camera-type with single lens, mirror and compound) arose independently, at least forty and maybe up to sixty-five times. Our old friend Richard Dawkins devoted a chapter in one of his books to 'the forty-fold path to enlightenment', emphasizing the repetitive ease with which natural selection could produce an eye, and so relieving you of the 'cold shudder' you experienced whenever you grappled with this problem.

There was no need to shudder and fear. The widespread sharing of regulatory genes and promoters between different types of eyes underlines the modular nature of both the genetic material and the developmental operations in which they are involved. Such modules switch partners through our turnover mechanisms and can be co-opted with relative ease into alternative networks of interaction. Indeed, the *Pax-6* gene itself, to take but one example, consists of three functionally important modules: the homeodomain, the 'paired' module and the 'octapeptide' module. Comparisons between species show that *Pax-6* genes contain various combinations of these three modules, not unlike the drawing of the plasminogens that I sent you some time ago. Transposition and gene conversion were no doubt involved in such shuffling, leading to the origin and spread of differently constructed *Pax-6* genes.

Natural selection would have had plenty to do at each branching point in the 'tree' of eyes, solving problems both of the external environment and of internal genetic turnover (molecular coevolution), but it would not have needed to go it alone in its pursuit of so-called 'improbable perfection'.

How do alliances shift?

In our current discussion of how all this might have taken place, we need to explain two aspects: the acquisition of a new subsection of a specific promoter and the origin of the ability of the *Hox* gene (or other developmental genes) in its new position to interact with new targets. The first event probably involves the transposition of a whole subsection of a promoter from one gene's promoter to another. So far, we have been dealing with simple modules of binding sites that are extensively shared between unrelated genes, such as *bicoid*-binding sites popping up in front of *hunchback* and at least eight other genes. These movements must have involved mechanisms of turnover, probably transposition, which can move one small piece of DNA from one position to another.

Similarly, whole subsections of promoters containing a combination of different types of binding site, in their different numbers, could jump from one gene to another. When this happens, it is important that the incoming subsection modules do not disrupt the normal functioning of the modules in the original promoter. There is a lot of evidence that modules and subsections of promoters are both independent and short range in their activities. So disturbance of the usual functions of a gene is kept to a minimum and new jobs can be assigned to the gene with its enlarged promoter.

The importance of these findings is that the regulatory regions of genes, and particularly of the genes involved in constructing different species, are very changeable. They consist of autonomous, free-floating and redundant modules, capable of moving from one gene to another and causing changes in patterns of individual development.

Let's now look at the second question: how developmental genes interact with a different set of target genes in the different species. This is harder to answer. I would guess that it is another consequence of molecular coevolution. Initially, there is a degree of relaxed and promiscuous cooperation between a developmental gene's DNA-binding domain (for example, a homeodomain) and the multiple variable binding sites in the promoters of the new target genes. Over time, this interaction would become refined through the action of selection-promoting compensatory changes in the homeodomains.

One last word and then I'll keep quiet (for a while!). It is interesting that quantitative changes probably precede qualitative changes in biology. This reminds me of Karl Marx's concept of dialectical materialism. Quantitative changes need to accumulate to a certain point before they cause a qualitative change in state – like heating liquid water to the fixed point at which it turns into gaseous steam, or cooling water to the point at which it turns into solid ice. This is similar to the processes of evolution, whether by natural selection or molecular drive. But biological organisms are not like the simple case of water because we do not know at what point we should be expecting a change of state, be it a new species, a new appendage or a new example of molecular coevolution.

With these generalized thoughts I sign off.

Dear Gabriel   Westminster Abbey

What can I say about your Hox! Hox! Hox! letter, but phew! phew! phew! Is there nothing sacred left in biology that cannot be teased apart and broken into its molecular constituents? What you have recounted seems to be 'reductionism' on a grand scale, by which I mean that ultimately there will be an explanation of all that moves in biology in terms of the genes.

I understand that you are definitely not subscribing to the view that there is a gene specifically for a given structure or behaviour; nor even that there are specific groups of genes for a given part of an individual. The evidence of the extensive sharing of modules that has taken place in the regulatory promoters of genes, or the multitude of functions for which a given gene can be responsible as it participates in a variety of genetic operations, should set the seal on any mistaken view that specific genes have evolved for specific functions.

Nevertheless, in your enthusiasm to explain the networks of genetic interactions that distinguish, say, a fly's wing from its haltere, it appears that you are succumbing to a view of genetic reductionism and genetic determinism. You seem to imply that genes rule the roost. No matter how inexplicably complex any given package of genetic interactions might get, by hook or by crook, the specific set of interactions under scrutiny has actually evolved, and that has inevitably involved the genes.

The extensive sharing of genes among your developmental 'packages' and the evidence that the gene can be defined only by its interactions (which I accept seriously undermines the argument that the gene itself is the target of selection) does not reduce the case you seem to have developed, perhaps inadvertently, for genetic determinism.

This seems to go against the grain of what you were saying in earlier letters. There you attacked the view that all of biology, including human behaviour, was explicable in terms of genes. Now, *in my own day, I was attacked, even without any knowledge of the genetic material, for suggesting that human evolution was no different in kind from the forces of natural selection that had shaped the forms and behaviours of all other living creatures. The attacks, naturally and understandably, came from the holders of religious beliefs that humans were exempt from the brute forces underlying animal diversity.*

I do not suppose that religious belief is in any way as strong in your times as it was in the pre-evolution times in which I lived. So I'm not referring to religion when I wonder aloud whether there is anything in biology that can be said to be out of reach of the genes. Surely you do not believe that human virtue or artistic appreciation or gentlemanly behaviour are as genetically determined as a fly's wing? Where exactly would you draw the line? I realize that my thinking is conditioned by the rather old-fashioned concepts of free will, culture and choice. But I would like to know whether you expect to find that packages of genes are responsible for some of these more obvious human characteristics, just as you describe the packages underlying insect body shape.

There is a danger, of course, that your insect examples could be dismissed as special cases having only to do with insects or other jointed animals. I would feel a lot easier in my mind if you could tell me what is going on in mammals, or even in that glorious pinnacle of evolution, the human brain. You hint that Hox *is involved, but you haven't yet sated my appetite. If you can get round to humans and convince me that the same genetic operations are involved, I'm ready to concede that Geoffroy was right all along. Who would have believed that poor old Geoffroy's view that a vertebrate is an inside-out version of an insect might have something going for it after all?*

Before you answer any more of my question, Gabriel, I want to clarify what I consider to be some important points. I sense that we are near the very crux of a new view of biology and I want to make sure that I'm not misunderstanding you. My concerns go beyond whether or not there can be evolution without natural selection. I think you have made it quite clear that without selection there can be no 'internal tolerance', as you say, to the continual restructuring of genetic material, as envisaged in molecular coevolution. There is no point in my denying that the genes and their regulatory regions are in a state of flux and that such turmoil cannot be restricted to being simply a new source of variation whose future is under the whim of selection. We have covered the two-step process of evolution in our earlier correspondence and you are right to consider genetic turnover as both a generator (step A) and spreader (step B) of new genetic processes.

What I want to do is to get a little deeper into what this means for biological functions. We've discussed this before in general terms. But now that you've given me the precise details of how developmental (and presumably behavioural) novelties arise, I can get my teeth into some real examples. I appreciate your running commentary on the bizarre ways in which genes interact with each other in the establishment of species-specific development. This helps me to understand your approach to the new findings in your day and age.

It's all in the mess

So let me begin. You have a tendency, which I do not fully agree with unless I've misunderstood you, to emphasize the 'mess' of developmental processes as if such a mess is not expected to result from evolution by natural selection. In other words, you point out that there is no simple 'hierarchy of command' in, say, D. melanogaster from the first maternally expressed genes such as bicoid through to the last expressed sets, or packages, of target genes. From your description of the system, and from even a cursory glance at your diagram of the network of genetic interactions, you are right to emphasize that there is no apparent logic to the system. Throw in a lot of 'on/off' control circuits and redundancy in switches and it would indeed seem that 'anything goes', as you say.

I understand that there might be views still prevalent in your day, as there were in my day, that there is a 'hierarchy'. Such views can lead to the belief that the genes at the top of the command structure are the least likely to change, for any change in them is likely to be catastrophic. On the other hand, changes in the last layer of genes to be switched 'on' and 'off' are more likely to be tolerated, as their effects would be kept to a minimum. We could view the hierarchy as an inverted pyramid of cards with the maternal genes at the bottom. Pull out bicoid *and the whole lot collapses. But your description of the complex networks suggests that the image of a pack of cards is way off the mark. The complete system is not likely to collapse on an evolutionary timescale no matter where we pull the gene from.*

What you have said about some of the differences in development between species means that new structures can arise through changes occurring anywhere in the network of interactions and not just in the target genes at the bottom of a hierarchy. So far, so good. What is observed is observed. But why cannot any of this arise by selection? Evolution via natural selection alone could give rise as much to an 'illogical mess' of genetic interactions as to a hierarchical pyramid of instructions. Evolution by natural selection is nothing but an opportunistic system. In a sense, we are back to our two-step process of evolution. Step A concerns the random production of mutations: that is, in the context of our present discussion, the random production of a new interaction (messy or not). By 'random' I mean that the new interaction is not directed, by any means whatsoever, to the adaptive requirements of the organism. If the new interaction proves to be useful in providing adaptive solutions to problems in the environment, then the novelty can be said to have spread by natural selection. We agree that selection is targeted on phenotypes, resulting from unique internal genetic interactions, and not on the individual genes themselves. But the random undirected changes in interactions, leading to variation in phenotypes, can be used opportunistically by selection.

If we continue with this blatant opportunism for long enough, we could well end up with the specific 'mess' of interactions in any given species of Drosophila. *In fact, with the aid of selection we could explain the five thousand different 'messes' of developmental interaction that we see in the five thousand different species of* Drosophila.

As you can see, I'm perfectly happy to accept your concept of a 'mess'. What I cannot yet gather from your remarks is the argument derived from this 'mess' that relates to adaptation. The 'mess', in all its complex webs of interactions, redundancies and modularities (TRAM), could be as 'adaptive' as anything else in evolution that may have arisen solely by natural selection. 'So what's new?' as some of your friends might say. I'm keen to hear from you about this.

Rather than use the details of development that you detailed in your recent letter, maybe we should discuss this issue on more neutral ground. Your question about the legs of the centipede seems an admirable way to proceed.

One hundred, but who's counting?

As we know, centipedes are arthropods – that is, jointed-legged animals without backbones. The same group includes the arachnids (spiders, scorpions, etc.), the insects and the crustaceans (crabs, barnacles, etc.). Centipede means 'hundred-legged', but of course this is not an exact figure, any more than the related group of millipedes has a thousand legs. Some species can have as many as 173 pairs of legs, one pair to each body segment, except the head segment and the last two posterior segments. Others have as few as fifteen pairs of legs. The average number is about thirty-five.

The main two questions in my mind that need answering are the usual interlocked pair of 'how?' and 'why?' How does the centipede acquire so many legged segments during development? And why does the centipede need to acquire so many legged segments?

I cannot answer the first question, although I suspect from what you have told me already about segments, wings, antennae and eyes, you'll have something interesting to say about legs. For the sake of discussion, I'll assume that there are sets of leg-forming target genes under the control of Hox genes and that the target genes can be shared and get into different interactions with many other developmental processes. Now, how can I explain centipede legs by natural selection acting alone? I think that I could do it, if you allowed me a bit of molecular drive arising from the usual flux in promoter binding sites – but that's cheating.

I suppose I would have to fall back on my notion of preadaptation. I think that we would both agree that centipede species with different numbers of legs are perfectly well adapted to particular habitats. I never really studied centipede adaptations, although I couldn't help seeing the things scurrying around under rocks. I was much more fascinated by beetles as a student, which I suppose *wasn't a bad choice given that they make up most insect species, and insects dominate the animal world. I remember being out in the countryside around Cambridge with two new beetle types, one in each hand, when, as luck would have it, I spotted yet another rare specimen. Without thinking I popped one of the beetles in my mouth while I went for the third. Never again would I sacrifice myself for science, as the imprisoned beetle squirted some intensely acrid fluid into my mouth! All three were lost.*

But let's think about centipedes, with each species seemingly happy with its lot in life, using its legs and much else besides to survive and reproduce in a given habitat. How did it all begin? Are 35 pairs of legs better for survival and reproduction than 34; are 36 better than 35; are 37 better than 36? And just to make it very difficult for ourselves, we also need to consider all the extra features of centipede legs, such as their joints, shapes, sizes, surface hairs, sensitivities to stimuli, flexibilities and speed of manipulation. Is each and every one of these features an evolved adaptation arising through natural selection?

Maybe you will allow me to simplify the problem by accepting that legs and all their attendant features come as complete modular packages, as you have explained is the case with other appendages. What I mean by this is that there is natural variation among a population of centipedes in their numbers of segments and legs, as complete packages. That way, we do not have to bother about natural selection slowly and gradually accumulating the genetic variants responsible for every hair, joint or shape of each leg. If we allow for this possibility of variation in complete packages, which I believe you would describe as a phenotypic module that can increase or decrease in number, then I suppose I'm free to suggest that perhaps 35 pairs are better than 34, 36 are better than 35 and so on.

This line of thought assumes that in a given environment 35 leads to an improved rate of reproduction than is the case for the bearers of 34. Then, once again, a change in circumstances merits the successful spread via selection of 36 over 35. And so on through all such steps until we reach centipedes with 173 pairs of legs.

I must admit that I'm already feeling a little apprehensive about this reply from what I have learned through your letters. I am assuming that environmental circumstances at each and every step lead to the gradual establishment of more and more pairs of legs. But since you asked me to explain it only by selection, I cannot escape from this fundamental assumption.

Are legs exaptations?

I do recall, however, that you once wrote about exaptations and linked them to neutral genetic drift. If I've understood this properly, you said that new mutations could spread and become established in a population because they initially involved neither benefits nor costs in the reproductive successes of their bearers. They would increase or decrease in a population because of accidents of sampling in constantly fluctuating populations of individuals or of sperm and eggs (I'll need to find your earlier letter to check whether I've got this straight.)

It could be that a population of centipedes with 34 pairs of legs was replaced by centipedes with 35 pairs of legs through neutral drift. Then, much later on, a change in the environment conferred some functional benefit on all these 35-legged creatures to the point where having 35 pairs of legs became essential for survival. In such circumstances, I agree with the two people you mentioned, Gould and Vrba, that such an acquired, co-opted, function cannot be called an adaptation because it didn't spread initially through natural selection. A functioning 35-paired phenotype needs another name, and exaptation seems to fit the bill quite adequately.

We've now moved from 34 pairs to 35 pairs using neutral drift; and 35 is an exaptation, by definition, and not an adaptation. Naturally, the contribution of the 35-paired phenotype to the survival of the animal might be the same in both cases. We cannot just look at the survival function of 35 pairs and decide from our observations whether it is an adaptation or an exaptation. I would even suggest that, if we measured the survival advantages of 35 pairs by cutting off some of the legs, we still could not say whether this is an adaptation arising only by natural selection or an exaptation having initially spread through neutral drift or some other sort of genetic hitch-hiking.

What are we now to say about the next step, 36 pairs over 35? Are we to assume that, just by the vagaries of sampling, a 36 becomes a new exaptive function? And are we to repeat the process through each and every one of the steps up to 173? I can see how your initial challenge to me to explain centipede legs by natural selection has only led me to another cul-de-sac of seeming improbabilities, even when I exploit the exaptive possibilities of neutral drift.

I could suggest, somewhat tongue in cheek, that we alternate our two scenarios: 34 to 35, selective adaptation; 35 to 36, neutral exaptation; 36 to 37, selective adaptation, and so on. I can't believe I'm getting so cynical even about my own concept of natural selection. Perhaps I'm subconsciously challenging you to provide an alternative explanation of centipede legs that is not equally improbable. I can guess by now that it will involve molecular drive, but even if it does, are we any better off? I'm keen to hear your response to this. The ball is in your court!

Back to Newton

As I told you recently, you can imagine my surprise, after talking with Sir Isaac who lies only a few yards away from me in the Abbey, to hear that physicists in your day are beginning to think that physics is no less an historical process than biology.

I've now heard the whole argument. What Isaac has been hearing from his physicist friends is that ...

Gabriel, I was just in the middle of expanding on my discussion with Newton when your next letter arrived. I've never heard you so excited. I remember you saying something once about a Drosophila with ten pairs of wings and eleven pairs of legs looking like the Manchester United football team, but I had no idea what you were alluding to. Now I can see what the fuss is about and why you have some justification as an evolutionary biologist in taking an interest in football. But first I had better re-read your latest letter before I continue with Newton. It doesn't matter to me that our letters are crossing in the post – as you say, there is nothing necessarily logical in biology or even in the discussing of biology!

Your obliging friend

TWO MINUTES THAT SHOOK THE WORLD

My Dear Charles

Please forgive me for being impatient and not waiting for your response to my last letter, but Manchester United, the most unpredictable and maddening of all football teams, has triumphantly won the European League football competition, at the same time as winning the English League Championship and the English FA Cup competition. I can already hear you wondering aloud what any of this has to do with biology. I beg your patience. First, I just want to relate the glory of the moment for no other reason than to triumph as a lifelong supporter of Manchester United. Second, there is a close resemblance between the game of football and the evolution of biology that I want to discuss with you. But for the moment that can wait. What I first want to relate is how a game of ninety minutes can be switched from a total disaster to maximum triumph within the last seconds. Not only was this particular game a case of unprecedented and unscripted high drama that caught the imagination of people all over the world, but it was an emotional melodrama that tested the hearts and nerves of all the supporters. Is there an evolutionary explanation for all of this? Genes for triumphalism? Genes for tribalism? Genes for play? Genes for obsession? Genes for escapism? Or is the explanation not in our genes at all – is it simply a cultural heritage of unadorned, shameless pleasure?

The first thing I want to say is that I have every right to support Manchester United because I was born there, the product of my absent Damascus-born father and my all-too-present very Mancunian Jewish mother. I have swum in the filthy Manchester

Ship Canal and spent half my teenage years waiting in the cold and relentless drizzle for buses that seemed never to come, but which when they did were full of heaving, coughing smokers. I've paid my dues at the gates of industrialism and earned my rights to glory – glory in the most adventurous football team in the world.

Charles, Manchester United is not a late-twentieth-century phenomenon, like most of the biology I'm telling you about. If you hadn't been so ill, especially in the last four years of your life down in the south of England, you could have attended the very first match of the team. Manchester, after all, is no more than 40 miles from your own origins in The Potteries. The first ever Manchester United footballers were the men of the Carriage and Wagon Department of Lancashire and Yorkshire Railway at Newton Heath in 1878, so the team was originally called Newton Heath LYR. No doubt you travelled on the LYR many times. These are humble beginnings for a team that has just played a game watched by millions of people scattered throughout the world, via our television networks.

The game in question was the final game of the championship series, to be played between the German team Bayern Munich and the English team Manchester United. I say 'German' and 'English' even though each team had many players from non-German and non-English countries. Such cosmopolitan mixtures in no way lessened the passion of the English for the 'English' team and the Germans for the 'German' team. Both teams had triumphed over the best teams in Europe during a gruelling season and the final game was to decide which was the best European team of the year.

If ever a game burst out of its boundaries to approach the proportions of myth, this was it. To understand the magnitude of what was achieved, you need to know only one thing about football: it is very difficult to score a goal. As its name suggests, the game is played with the feet, which is a considerable skill appreciated by all people around the globe – except in the United States, where people don't understand why the ball isn't picked up and carried in the players' hands.

On average, only one or two goals are scored in a match – say, one goal every seventy-five minutes. At the level of the European Championship, one goal could decide the outcome, and one slightly dubious goal scored in the first six minutes by Bayern Munich seemed to have settled the matter for the rest of the game. Manchester United fought like heroes, but there seemed to be no way through the systematic defensive tactics of the Germans. It was Stalingrad in reverse. As the dying minutes of the game approached, the match officials were tying the Bayern ribbons to the winners' cup and the Bayern players were waving to their supporters. And then, in a brief moment of unprecedented drama, the world came crashing down around their heads.

Three minutes were added to the end of the game for time lost in attending to players' injuries. United had already put on two fresh players as substitutes. At 90 minutes and 36 seconds, the screams and whistles among the crowd of ninety thousand people in the huge Barcelona stadium were deafening. They were designed to tell the referee, as if he didn't know, that he should be thinking of blowing his whistle for the end of the game. Some of Manchester United's star players of earlier years had already left the stadium, unable to witness the impending defeat of their team in the most important game any club could play. Then the first United goal came. A high ball from Beckham, United's star player, into the German goal area, some confusion, a pass out to a United player, a pass back towards the goal to another United player, the first substitute, and then the ball was threaded through the eye of the storm into the German goal. Everyone, United and Bayern players and their supporters, was stunned – a goal in injury time, a heroic reprieve from defeat. Less than two minutes left and then both teams would have to play for an additional thirty minutes to break the deadlock.

But the extra thirty minutes never came. At 92 minutes and 4 seconds, another beautifully weighted cross from Beckham floated towards the Munich goal. The ball was headed on by the first substitute towards the second substitute, who with an elastic outstretched leg popped the ball into the roof of the net. A kick, a header, a kick, a goal; history made! Two goals in less than two minutes in injury time. Everyone crying, English and Germans; Munich players being pulled to their feet by the referee to restart the match for the last fifty-six seconds.

Oh, the unbearable darkness of defeat for men prematurely celebrating their triumph; and the unbelievable headiness of triumph for men previously down in the muckpile of disaster. Like history, sport is the story of winners. But there are two sides to every coin and without the pathos of the losers there could be no heroism for the winners; without catastrophe there can be no triumph.

And they did it for me, not for the millions of other supporters. I am not millions of other supporters – I am my own uniquely developed individual, intimately relating, kick by kick, to the bewildering patterns of ball passing among eleven other unique individuals. Such identification cannot be genetically determined because the chance of my DNA fingerprinting patterns matching those of the players is too low to consider. It is not nationalism because the team is cosmopolitan, and besides, my grandparents hail from three different continents. It is not tribal because my life in academia is too remote from the riches and glamour of their lives. But somewhere in my genetic and environmentally induced development, a unique combination of molecular interactions took place that led me to revel in the high-risk glamour and attacking skills that uniquely define the Manchester United team. But I'm not forced or predetermined to revel; it is my individual free choice. I can stop at any point. And millions of other individuals around the world developed into similarly inspired, freely exercised identities, but not, I wager, because of the same set of genetic and environmental interactions that made me. As unique emergent individuals, not as average members of the same species, their particular personalities took shape as a result of unique sets of interactions. There are no genes for the appreciation of glamour; no genes for the envy of sporting skills; no genes for the harmless emotional responses to triumph and defeat. Our behaviour is not fashioned by the assertiveness of specific genes for specific jobs, as they pursued their selfish interests through the antics of our forebears, half a million years ago.

Our responses to sport are individually based; they should not be assumed to be universal features of the human race, implying an underlying genetic determinism. When millions of individuals share similar responses to a football team, it is because they choose, individual after individual, to be influenced by the cultural and social habits of the day. The speed and intensity of such influences and our freedom to choose at each and every singular moment, frankly leave the genes in their starting blocks.

If this can be said about sport, then it can also be said about our other human attributes, for better or worse. Are the groups of marauding individuals who are murdering, raping and pillaging their way across the Balkans re-enacting the supposedly genetically determined hunter-gatherers as they supposedly adapted to their own harsh environments? Or are we humans capable of entering, sometimes willingly, into the most inhuman of crimes by the manipulative techniques of modern 'culture'? We are each capable of responding as much to evil suggestion as to Beethoven and Beefheart. The unique sets of genetic interactions in each of us give rise to multifunctional and multiresponsive individual minds and bodies. Our biology makes us free and ultimately responsible. We cannot blame individual genes, selfish or ignorant, for the Swiss Army knife of our individual human psychology and responses.

The average human hand is built by genes during development, but no exact predictions can be made about any particular hand. Similarly, and more importantly, there is no genetic determinism over the uses to which we decide to put our hands: are there genes for writing, for scratching, for gesticulating, for making tools? Any notion of 'the function of the hand' shaped by hand-determining genes is vacuous at best, and reveals an ignorance of modern genetics at worst.

Similarly, our average human brain and consciouness emerge from an unfolding of interactions between multifunctional genes, most of which will have TRAM characteristics. But this does not imply that our individual brain-based psychology (self-consciousness, aesthetic appreciation, humour and much else that is currently wrapped in the flag of evolutionary psychology) is determined by the victors of the wars of genes waged hundreds of thousands of generations ago.

Were you, Mr Darwin, to be living in my time, I'm sure you would find it as difficult to adjust to our psychologies and behaviours as I would to yours were I transported back to the nineteenth century. No matter what universally shared evolutionary influences there might be in the genetic make-up of you or me as individuals, we uniquely reflect our different worlds. If you had been born when I was born and had followed my life's trajectory, you could very well have chosen to end up a Man. U. supporter, tears streaming down your face. Had I lived with the means and peace that were at your disposal, I might have been a courteous and gentlemanly father figure to all around me.

Each of us has the genetic make-up to respond appropriately as unique individuals to different circumstances. Our genes are in a constant state of flux on an evolutionary timescale, and our developing phenotypes are in a state of flux in the span of a human lifetime. Neither evolution nor development is fixed and locked into predetermined paths. In biology, $1 + 1 = 7$, or 9 or 13. We are an interactive, combinatorial system. There is an unknowability and unpredictability of DNA that needs to be in the forefront of our minds before we make naive statements about genetic determinants winging their selfish way through evolution.

Football versus chess

Well, it looks as if I've strayed a long way from football. But I haven't; I've come full circle. I believe that football is a much better metaphor for biology than, say, the game of chess. Both games have rules, but there is a fundamental distinction between the two. In football, the rules simply specify, by and large, what you are not allowed to do. Within the confines of these rules, the players can do what they want. They can all stand in a line in front of their own goal or they can all stand in a circle in the middle, if they wish, although the team wouldn't win many games that way. But in real games, players can interact with and pass the ball to each other in an infinite number of permutations. A given player can be involved both in a defensive exchange of passes and in an attacking exchange of passes. Within the 'don't do' rules, the art of the game is the generation of unexpected novelty in the patterns of passing.

In chess, the rules impose a rigidity and fixity over the game unlike football. Each different piece is fixed for the types of move it can make and in the positions it can occupy on the board. It is as if a player is only allowed to kick with his left foot, or only run three paces in a diagonal direction before passing. True, a game of chess is full of constantly changing patterns, but the available variety of patterns at any given moment is limited by the rules of what each different chess piece can do. That is why the opening moves of chess often follow predictable patterns. There is a hierarchy of power among the pieces, from pawns to royalty, that essentially determines the outcome of the game. The winner is the one who can use such fixed powers to greater effect, relative to the fixed powers of the opponents' pieces. In football, the differences between the teams are the result of how individual players have uniquely developed during their time on Earth. Within the confines of the 'don't do' rules, a wide-ranging combination of ball passing can take place on a 'can do' basis. In the same way, our hand and brain functions 'can do' in a seemingly unlimited manner.

Biology is more like football than chess. The 'rules' of biology are a reflection of past historical events – don't produce a head in the middle of the abdomen because you happen to be on a branch of the tree of life where such things never happened. If you want to survive and reproduce, don't do that. But within these historically determined limits, there is no end to the variety of permutations that our redundant and promiscuous genetic units can engage in. A major criterion for success is that a given combination of genetic 'passes' produces its 'goal' – an individual capable of developing and reproducing while maintaining full compatibility with similar individuals with whom it will eventually have sex.

Naturally, I cannot stretch the football analogy too far to embrace molecular coevolution. Unlike biology, as one great British coach said, 'football is not a matter of life and death – it is more important than that.' My main point, which I think you will appreciate, Charles, is that our view of evolution has changed from one in which each and every gene has acquired a fixed selective power, specifically placing it in a hierarchy of commands during development, to one in which the gene–gene interaction and the flexibility and promiscuity of such interactions builds unique, unpredictable individuals. Our multifunctional hands and brains could be as much the inevitable outcome of how our bodies get built as the purposeful adaptive wonders of long-forgotten solutions to past ecological problems.

The gene has no meaning and no selective powers outside of its interactions. The gene is ignorant about what is being done to it individually and about what particular evolutionary path it might follow. Only the self-reproducing individual has an evolutionary meaning. Long live football and long live chess. Long live the ephemeral phenotype and long live genetics. But the gene as a solitary unit, I'm sorry to say, is not capable of winning any game, football or chess, over any timespan.

You can see, Charles, what the recent triumph of Manchester United has done to me. Non-genetically determined euphoria has led me into heady analogies between football and biology. But as the great writer Samuel Beckett said in his 'sucking stones' metaphor for the dry pursuits of philosophy, 'Deep down I didn't give a tinker's curse whether I sucked a different stone each time or always the same stone until the end of time.' Metaphors are of limited use. Did you by any chance go to watch the Newton Heathens of the Lancashire and Yorkshire Railway company?

Glory, glory Man. United!

Gabriel

Westminster Abbey

My dear Gabriel

Your enthusiasm for Manchester United is infectious; I can understand now why some of my compatriots believe that God is English! It is, indeed, a pity that I didn't take the opportunity to watch the Newton Heathens – not that this would have been easy, for Emma would certainly not have allowed me out of her sight in 1878 and beyond. Remember that I only had another four years to live and they turned out to be physically very miserable for me. Still, the opportunity was there to have created a direct link with you, a full century later, through your beloved Mancunians.

Your contrast between chess and football is instructive and does underline a lot of what you have been saying to me. I must warn you, however, if you haven't already spotted the dangers for yourself, that we can no more say that 'anything goes' in football than we can say the same in biology. There might, indeed, be nothing in the rules of football that defines a winning move or prevents a whole team standing in a line in front of the goal. But many such possibilities are hardly going to win games. Success in football, like chess, surely relies on the accumulated experiences of successful moves in former games. Despite the lack of rules limiting what each player can do, the wisdom of history must permeate football, just as it permeates the game of chess. The non-permissive rules, if I can phrase them that way, are as pervasive in football as they are in chess; it's just that they are not written down.

Nevertheless, if I set that point aside and get into the spirit of what you say about football, I can see that there is a freedom to create an unexpected combination of moves between free-wheeling players that cannot arise as liberally among chess pieces, with their fixed instructions. In that sense, there is some link between football and biology: an unexpected set of interactions between multi-functional modules (I'm getting quite into the swing of your phraseology) can produce a most unusual goal or new biological structure (multiple-legged segment in centipedes!). Both the goal and the structure exploit existing flexible modules and the wisdom of past interactive experiences. I can even understand how the suddenness and finality of the seeming miracle of Manchester United's two goals in the last two minutes of the Championship elicits the same wonder that we biologists have for the bewilderingly unexpected variety of life's creatures.

Universal selection?

After this enjoyable little chat about football, I hope you don't mind my weaning you back to more serious issues. I was just going to tell you about my very interesting conversations with Isaac Newton when your last letter arrived. Apparently, Newton has been in regular correspondence with a number of physicists and is very well informed of new developments in your late-twentieth-century understanding of the fundamental building blocks and forces that govern everything from the very smallest, what he calls the sub-atomic particles, to the very largest, the cosmos itself.

What Newton has been telling me relates to our earlier discussions about the laws that govern physics and chemistry, and the lack of such laws to govern biological evolution. We both assumed, I presume, that the laws of physics are fixed and eternal, independent of the history of the Universe, while biology, in all its forms, is a product of historical events or accidents that are never to be repeated. I drew your attention to the distinction between a population of water molecules and a population of organisms. Water molecules behave in regular ways as the temperature and pressure are raised or lowered, irrespective of time and place. But in organisms, whatever events might be occurring at a given time or place cannot lead to the formulation of laws to predict what will happen at the next moment in time or place. This is because of continual changes in both the genetic composition of the individuals in each new population and the composition of the local environment. There seems to be no regularity of events in biology on which hard-and-fast laws can be formulated, as we understand them to operate among inanimate objects. I attempted to turn 'gradualism' and 'the principles governing natural selection' into biological laws, but I notice that you've kept unusually quiet about that.

Do we really understand the laws of physics?

The problem seems to be as follows: almost all the matter in the Universe is made up of four kinds of elementary particles – protons, neutrons, electrons and neutrinos – which interact with each other through four basic forces – gravity, electromagnetism and the 'strong' and 'weak' forces in the nuclei of chemical elements. Apparently, about twenty different parameters or variables define the behaviour of these particles and forces. For example, particles can vary in their mass and in the strength of their electrical charges, to take just two parameters.

From what I'm hearing from Newton, there are ideas in circulation, particularly from a physicist called Lee Smolin, that the laws governing the current Universe might result from historical processes, no different in principle from the historical events that shape biological evolution. In a nutshell, the variety of parameters that describe the four basic subatomic particles and the four basic forces, present throughout the cosmos, could be viewed as the result of natural selection operating among alternative universes in which the parameters determining the behaviour of particles and forces were once different.

Apparently, the belief that the parameters can change with time emerged from the unlikelihood of arriving at our current laws of physics by chance. This is the same problem we had in trying to create a Boeing 747 from a tornado in a junkyard.

The sheer improbability of creating stars and stable nuclei from the available particles and forces depends on the very narrow ranges over which the current parameters operate. For example, calculations have shown that our Universe could only exist in one tiny spot in the two-dimensional space defined, say, by the parameter of mass, as measured in a neutron and an electron. A plot of neutron mass against electron mass defines a largely unoccupied space. The vast majority of matches of a given neutron mass to a given electron mass, other than the actual neutron and electron masses already existing in the Universe, would give rise to unstable nuclei. Without stable nuclei there are no stars and without stars there is no biological life. If this procedure is repeated using all twenty variable parameters in the construction of a 20×20 dimensional space, then our Universe occupies an even smaller, highly improbable spot of space.

Newton tells me that his 'gravitational constant', a number used to measure the gravitational force between two bodies of given masses, is as small as 10^{-38} between two protons. As you know, this is a 1 separated from a preceding decimal point by 37 zeros. This is a ridiculously small number, but if it got stronger by, say, dropping one zero, then the lifetime of a typical star would fall dramatically from about ten billion years to ten million years, making the evolution of life as we know it impossible.

According to the idea that the masses of the four elementary particles are all set independently of each other, and also that the strengths of the four basic forces are independently variable, getting conditions just right for the existence of stable nuclei, chemical elements and everything else in the Universe as it currently exists is frighteningly difficult to achieve by chance. It's as if there are multiparameter knobs on a switchboard for each of the particles and forces, and these have to be set by chance to very tight specifications before we can explain ourselves and our Universe. Change any knob by a small amount and the Universe around us will disappear and a different universe will exist. Apparently, the calculations show that there is only one chance in 10^{229} – one chance in a 1 followed by 229 zeros – of the current Universe existing. It is impossible to imagine the sheer size of this number unless we compare it with the calculated number of stars we can see from Earth (about 10^{22}), which contain about 10^{80} protons.

So how did the Universe arise? God is one possibility. Newton used to think that this was the ultimate explanation, but he realizes that most of your twentieth-century scientists will not entertain any such mysticism. Another possibility is that there were 10^{229} universes, of which one occupied, by chance, the exact parameter space required of our current Universe. Newton tells me that Smolin doesn't like this explanation because it's not rational: it can explain anything. I'm not sure what is meant by that, but Smolin suggests that if such an irrational assumption had been applied to biology, I would never have come up with the idea of natural selection.

So how are these poor physicists going to solve their little problem? If there is something like a natural selection of parameters going on, where does the variability in parameters come from? Where are all these alternative points in parameter space, defining alternative universes, for selection to get its teeth into? Unless the physicist can show that the twenty parameters can actually change over time, there is no grist for the selection mill.

I don't want to get too distracted by the amazing world of Big Bangs, black holes, string theory and everything else that makes Newton jump up and down the aisle. The basic argument for natural selection in the Universe is that it is possible for parts of our Universe to bounce into alternative universes, with different knob settings, every time a black hole is formed. A black hole is a region of space and time where gravity is so strong that nothing can escape from its gravitational pull, not even light. It is believed, I'm told, that they form after the collapse of a massive star when it runs out of the energy generated by its nuclei and can no longer support its own weight. There could be as many as 10^{18} black holes generated in our Universe.

Now comes the trick. Smolin believes that the physics of small particles suggests that black holes are not singular points reflecting the end of time and space, but can emerge on the other side into alternative universes, much like Alice through the looking glass.

This could happen if the collapsing star exploded once it reached a very dense state and formed different universes with different parameter settings out of the debris. Presumably this debris had decoupled the elementary particles or forces from the parameter settings with which it entered the black hole.

The point Smolin is making to Newton is that some black holes will throw up more offspring universes than others. In some of these offspring universes, the parameter settings will be such that even more black holes can be formed subsequently, leading ultimately to even higher numbers of alternative universes. Eventually, the parameters would reach settings of our actual Universe with its 10^{18} black holes. So our Universe arose from the inevitable selection of those regions of parameter space that were more likely, through the chance settings of the knobs, to spawn many universes, some of which would have even greater potential to spawn alternative universes.

I think you will agree with me, Gabriel, that this is an example of natural selection at work. The so-called Big Bang could have been a black hole spawning multiple universes, which were refined by selection to produce the current Universe. The total lifespan of the Universe, a dozen billion years, is long enough to explain the origins of chemical elements and life itself.

If all of this is true, it should be possible to explain our Universe, as well as life, in terms of history rather than general principles. I don't think we would disagree about this because we still believe that the evolution of biology is a lawless, historically influenced and historically constrained process. We can take comfort from the fact that it might not, in these circumstances, be any different from the lack of principles governing the origins of our Universe.

Life and the cosmos are not unimaginable and improbable events, once we take on board the inherent potential in both systems to generate variable 'offspring' to differing extents. We could even go as far as to say that the chance of generating some evolutionary paths through the multi-dimensional space of genotypes is 1, given the incessant drive to diversify in biology. Similarly, the probability of arriving at our Universe with its ability to create black holes and new universes is unmystifingly good, given the inherent flux in the requisite parameters.

It's all beginning to look too easy! By the way, Isaac passes on his regards to you.

Your physicist servant!

BORN TO ADOPT

Dear Charles

I was in Westminster Abbey yesterday, and I do believe I heard you and Newton chatting away! I took my two young sons Noam (13) and Alexis (9) up to London for the day. They have been so involved in one way or another with our correspondence that they were quite excited at seeking out your tomb in the Abbey. I'm ashamed to say that I've never been to visit you before. I was expecting some sort of Victorian extravaganza with suitable Latin inscriptions from the Psalms. I was surprised to find the plain white unadorned flagstone, devoid of words, except for your name and lifespan. Your tomb is more like some memorial to an unknown soldier than the last resting place of the person many hail, at the end of the twentieth century, as the Man of the Millennium. I think the simplicity of the inscription and failure to say what you were doing speaks volumes for the ambiguities and tensions among the religious and scientific establishments at the time. By contrast, Newton's tomb is veritably throttled with inscriptions and wise words, in an elaborate construction full of ornamentations and embellishments, and topped with his noble features as adoring cherubs pander to his every whim.

The hidden drama of Westminster Abbey hosting the remains of a man who tore the heart out of religion must be one of the great ironies of the clash of ideas. Whichever way we turn there is no escape from the godless explanation of the origin of life and of humans by natural processes, explicable by the scientific method. This does not mean that people had to abandon their belief in the supernatural, but that the central mystery of our origins, which had fed religious belief for tens of millennia, had finally been superseded. After your noble efforts, there could be no turning back from the scientific evidence that the world and all it contains can be explained in natural terms. The myriad forms of life around us carry in their structures and activities the history of their own making.

What I like about your conversations with Newton is that there seems to be a history of events surrounding the origin of the Universe that is similar to the events surrounding the origin of life. And this can be derived from the current structures and activities of the fundamental particles and forces. We seem to be entering a world of understanding in which there are no absolute entities, fixed and eternal, that govern our past, present and future. The removal of the supernatural as a fixed controlling entity, existing for all time, was the first step in this direction. I understand that this was not necessarily your intention in advocating our evolutionary origins, but you did admit to feeling like you were 'confessing to a murder' when you published *On the Origin of Species*. If Smolin and other contemporary physicists are right, even the parameters that define the structures and strengths of particles and forces were not necessarily fixed and immutable from the Big Bang.

Personally, I would not, as you might expect, describe the historical unfolding of the cosmos, if that is what it was, as the result only of natural selection. Smolin fixes on natural selection because that is all he seems to know about, alas. I'm sure, if we put our minds to it, that a contribution from neutral drift could also be entertained, as indeed could a contribution from molecular drive. Not in terms of genes, of course, but in terms of accidents of sampling in the former case, and inherent turnover and differential spreading of variable cosmological units in the latter case. I would argue that Smolin's picture of the origin of our Universe, as you have described it, is closer to molecular drive than to natural selection. He does not provide any particular adaptive reasons why one set of parameters is preferred over another, except that one set of parameters spawns more universes than another. As we discussed, natural selection is not about differential reproduction in itself. It is about differential reproduction as an outcome of differential abilities to produce solutions (adaptations) to external problems. Smolin's case, on the other hand, rests on inherent differences in spawning more universes as a property of the parameter settings. This is equivalent to the case of biased gene conversion, in which the sequence of nucleotides in a given gene can affect the direction of conversion.

Looking for the eternal

Be that as it may, the important issue is that there seems to be nothing in our understanding of biology or physics that supports the idea of eternally fixed, independent entities lying at the heart of the natural world. This is as true of the genes as it might be true of the fundamental particles and forces. The genes and the particles are not the fixed, free-wheeling masters of the Universe.

There is an interesting link, which I will mention but not develop, between militant atheism and the false notion of the eternal gene. The belief in the eternal gene cannot be sustained alongside a belief in an eternalized 'God'. It requires the belief in the supernatural to be relentlessly destroyed and removed from all rational minds. The vehemence with which this anti-religious war is waged, by well-known advocates such as Richard Dawkins, must be connected with the belief in the eternalized gene. There seems to be as much religious fervour devoted to the false notion of the self-replicating gene as there is to the earlier notion of an eternally operating, supernatural originator of our natural world. One misconception has simply replaced another. And as with all religions, there has to be a head of the Church and a source of received and exclusive wisdom. And this, you may be shocked to hear, is your true self and your process of natural selection.

The following quote, you might agree, is the nearest we get in science to the transformation of a working, historical process into a religious non-scientific mantra: 'The theory of evolution by cumulative natural selection is the only theory we know of that is, in principle, capable of explaining the existence of organised complexity. Even if the evidence did not favour it, it would still be the best theory available' (Richard Dawkins, *The Blind Watchmaker*). I love that bit 'even if the evidence did not favour it'. No wonder there is no room for the other concept where no evidence is needed, that of 'God'.

My own views on 'God' are more prosaic and pragmatic. I do not need to reject the idea of supernatural influences in all natural things in order to bolster my acceptance and propagation of the scientific method. That methodology is quite capable of standing on its own.

Indeed, the widespread need for humans to believe in the super-natural is a legitimate target of scientific enquiry. What was it in the evolution of humans that led to our overheated brains becoming so relentlessly self-conscious, forever pained about our past, present and future lives? This simple question needs to be answered if we are to understand the recent decision of the State of Kansas, USA, not to allow your theory of evolution to be taught in science lessons in its schools. They consider it unproven, unlike the Old Testament creation story that can now be taught as science. This is intolerance and child-manipulation on a vast scale. But our job, as scientists, is to investigate what it is that the legislators fear. Why do only 10 per cent of Americans believe that 'God' played no role in evolution?

Gerry Edelman has spoken about the simplicity of animal consciousness living in the 'eternal present' while humans suffer the perception of times past and times future. Does a gorilla, watching the sunset after a good day's sex and foraging, think that he has had a good day and hope that tomorrow will be the same? Gorillas probably don't, but humans certainly do. So if some humans want to find solace in the eternal supernatural, or in the eternally fixed entities of particles and genes, there is no harm in that, provided it doesn't interfere with the pursuit and dissemination of science as a natural phenomenon.

My own toying with the mystical came in three doses. The first was at the age of nine or ten when I was momentarily laid up in bed, while all my friends were out roaming the countryside. I remember throwing thousands upon thousands of kisses to heaven. I couldn't stop my lips blowing kisses. It was a secret communion, for reasons totally unknown, between myself and what I believed at that time existed 'up there'. That episode has not affected me in any way. It has not spurred me on to seek deeper spiritual sustenance. I remember it now only in the context of this letter, decades after the event. Was it a random accidental set of firings in the neurons of my brain? Who knows? What it was, it was.

The second episode was more of a real test. Yom Kippur is the holiest of holy days in the Jewish calendar. All men and women over the age of thirteen have to fast to atone for their sins in the preceding year. I had just turned thirteen and the dilemma was upon me: should I fast or not? Before then, when there was no need to fast, I had fasted with the best of them. But now I was obliged to fast – it had been decreed 'on high'. So I went with a couple of friends across the road from the synagogue to buy three cream buns – non-kosher ones at that. All three of us were determined to eat. I opened my mouth, the bun was at the lip of the orifice, the eyes of my family were on me from across the road and... I paused. For one real, intense microsecond, which I remember to this day, I expected a clap of thunder and a bolt of lightning to wipe me off the face of the Earth. It didn't happen and the rest is history. My premonition of instant obliteration was not unlike the day when the penetrating mind of Francis Crick demanded that I prove to him that without an evolutionary perspective there is no understanding of development. I feel relieved at passing both tests!

Finally, I remember the summer I finished writing my PhD thesis on the genetic control of chromosome behaviour in wheat. I had spent three years on this research. I was obsessed with my experiments and intensely filled my time investigating all that there was to know about the life of plants. I was a plant too – my roots went deep into the earth, watered by the font of hard knowledge. And in a moment of supreme arrogance, which later turned out to be the depths of ignorance, I stood alone on a summer's evening looking out on a Cambridgeshire wheat field, saying to myself 'I understand all of this.' I was the world's expert. I knew the whys and the wherefores of every wheat gene as it governed the Mendelian and non-Mendelian behaviour of the chromosomes. I knew the path of every hormone and the activity of every molecule, as plants made their contribution to life on Earth. The feeling bordered on the mystical, but it was all baloney. I no more understood the genes and the molecules in those pre-molecular days than I understood why I felt that way. My understanding today, thirty years later, is a little better, but I want to meet the scientist who claims that he or she understands it all and who arrogantly turns the supposed completeness of scientific knowledge into a 'religious' belief or comfort.

All science is preliminary and we have to recognize and make the best of our ignorance. Scientific arrogance should not be used as a weapon against the supernatural. So used it simply demeans the honesty and acceptance of the transitory nature of knowledge, which is at the heart of science. I know that physics has moved on from

Newton and that biology has moved on from Darwin, but we are all, Charles, still scratching at the beast produced by evolution.

With that in mind, let me get down to some of your questions.

Pulling one's leg

I think that the best way to answer many of your general questions about genetic determinism, 'illogical' genetic networks and whether the centipede is an adaptation, exaptation or, dare I say it, an adoptation is to get down to explaining some more modern biology.

If we are going to focus on centipede legs, let me tell you the story of legs as it is known so far, first in invertebrates and then in vertebrates. All legs are long drawn-out structures. To understand how a leg develops in *Drosophila*, you have to imagine a flat circular area in the middle of the leg segment of the larva filled with concentric circles, just as if a radio aerial had been pushed down to its shortest length. Pull the aerial from the smallest, innermost ring and it can extend to form the farthest region along a near–far axis. There is no need for me to go into details of the genetics of legs. As you might predict by now, there are a number of interactive genes, some of which are also involved in forming other appendages such as wings. There is one particular gene called *distal-less* that is most active in the innermost circle of our set of concentric circles. The reason I mention this is that the *distal-less* protein is also within the innermost circle of 'eyespots' on butterfly wings. Hence, by a roundabout route, I've brought you to the genetics of 'eyespots' as promised. Once again we see that a small number of interacting genes, used to fix the precise positions of 'eyespots', are shared by other developing appendages.

Before I look into some of the evolutionary implications of leg development as it relates to our centipede discussion, I think it is about time that I satisfied the curiosity you expressed about vertebrates.

The first thing to know is that the cluster of eight *Hox* genes that we saw in *Drosophila* exists in all examined multicellular organisms, including ourselves. Not in the same numbers, of course, but interestingly in the same order on the chromosome. In mammals there are four sets of clusters, with up to thirteen genes in each set, although within each set one or more of the genes can be missing. As

I explained in my letter about species formation, some fish species seem to hold the record for the number of different sets of *Hox* clusters.

Same genes; same story; different outcomes

The story of the genes controlling the development of legs and wings in chickens is a good example of what is happening in vertebrates. Chickens, like humans, have four sets of *Hox* clusters called A, B, C and D. The distribution of *Hox* proteins from some of the genes in clusters A and D determines the positioning of the bone structures in the developing wing. As in *Drosophila*, the expression of the *Hox* genes is controlled through the combining activities of a number of other genes, including the vertebrate equivalents of *engrailed* and *hedgehog* (now called *Sonic Hedgehog*, as most children will appreciate), *wingless* and so on. There are also some genes, probably recruited from other activities, that are specifically involved in vertebrate limb formation. But the overall message is the same as in *Drosophila*.

The establishment of new genetic interactions involves the successful recognition of regulatory proteins of their respective multi-copy and modular binding sites within the compound promoters of their target genes. As an example, new relationships between master genes and 'packages' of target genes have occurred to ensure that arms or wings (forelimbs) become different from legs (hindlimbs). The decision whether to turn a developing limb bud into a wing or a leg depends on two regulatory genes called *Tbx5* and *Tbx4*. *Tbx5* is expressed in the forelimb bud and *Tbx4* in the hindlimb bud. Now, using all the tricks of modern genetic manipulation, the leg can be made to look more like a wing, in terms of its bone structure, if *Tbx5* is artificially expressed in the hindlimb bud. Similarly, if *Tbx4* is expressed in the forelimb, where it wouldn't normally be expressed, then leg-like bone structures form instead of wings. By such simple means, legs can switch to wings and back again.

This is not, of course, the whole story, as you might guess, but it illustrates that the 'package' of target genes needed for, say, wing development exists in the hindlimb bud and can be sought out by the wing regulatory protein *Tbx5*. Many of the genes used to set up the axes of the hindlimb and forelimb are shared by the two limbs, as indeed would be many of the genes in the 'wing' and 'leg' packages of target genes. We have come across similar extensive sharing in my description of the wing and haltere differences under the control of the *Hox* gene *Ubx* in *Drosophila*. The difference between vertebrate wings and legs depends on differences in the combinations of many shared genes.

I cannot resist giving you one more example of important differences that arise from promoter restructuring. The backbone of the mouse consists of seven cervical (neck) and thirteen thoracic (chest) vertebrae, whereas the chicken has fourteen cervical and seven thoracic vertebrae. The different number of cervical and thoracic vertebrae in the two species is reflected in the different extent to which two particular *Hox* genes of set C are expressed in the developing embryo.

Now, the promoter of one of the *Hox* genes (*Hox* gene number 8 in set C) consists of a number of redundant sites involved with the response of the gene to regulatory proteins. Differences have been found in these promoters between the mouse and the chicken, and also between the mouse and the whale. Using the usual genetic tricks, these differences can be shown to be responsible for differences in the patterns of activity of the gene along the developing backbone of the mouse, chicken and whale. Naturally, there is also a good deal of conservation of binding sites in the gene among these species. As with many other *Hox* genes, it is crucially important for determining structures along the backbone in adult vertebrates. But there are subtle differences in promoter composition that are linked to the large differences in the types and numbers of vertebrae in backbones of different species.

Let me return, however, to the legs of arthropods. The genes involved with leg formation in butterflies, crustacea, grasshoppers, spring-tails and centipedes involve the same *Hox* and other genes that we saw in *Drosophila*. But there is no logic to how they are involved with legs as we move from one species to another. Getting a leg in a particular segment of a particular species depends on patterns of 'on' and 'off' switching of genes. Such patterns differ significantly between species.

During the evolution of the arthropods, changes in the regions of activity of master genes, such as *Hox*, coupled with changes in the pattern of interactions among the *Hox* genes and between *Hox* genes and subsets of target genes, can explain the huge variety of forms of segmented animals. Go to any sea-food restaurant and you will see what I mean. All of this variety is the result of new combinations of regulatory master genes and their binding sites in the promoters of target genes, exhibiting the usual features of TRAM systems. An inherent and inescapable flux in promoters has an important knock-on effect on the initiation and spread of new body shapes.

Swopping regulatory circuits: simple to do, profound in effect

A cell's physical make-up is determined by the particular combination of genetic circuits taking place during its development. Change the links between the master genes and the target genes during evolution, and new cell phenotypes emerge. In the words of Garcia-Bellido, in a classic paper written in 1975, 'The appearance of new selector genes does not demand new realizator [target] genes, but only a qualitatively different utilisation of those already existing, so that, in this sense, the amount of genetic information for evolutionary complication is kept to a minimum.'

There are many good reasons to retain the idea of master genes calling on different subsets of target genes as evolution unfolds. The conservation of the *Hox* genes alone across all multicellular animals, dating back some thousand million years, is proof enough of their longevity and necessity. As we saw in the case of the eye-controlling gene *Pax-6*, a master gene can be swopped around and still carry out its function in species of vertebrates and invertebrates because it is capable of interacting with different packages of eye-forming target genes. The similarity of genetic circuitry between legs, wings and 'eyespots' suggests that eyespots resulted from the use of an existing circuit of interactions in cells that would not normally express this circuit. What we are seeing now is a recycling of whole circuits of genetic interactions to engage in new processes.

Similar use of a self-contained circuit of regulatory interactions has been made in the development of the egg, the determination of neuroectoderm, the emergence of sensory organs and segmentation.

Making *Hox* more subtle

One of the problems with accepting that *Hox* master genes play an important role in development has to do with the 'hopeless monster' syndrome. If master genes could be observed as a little less dictatorial and a little more open to outside influences, perhaps they could be more acceptable to natural selection.

Michael Akam of the University of Cambridge points out that the *Hox* gene *Ubx* can indeed act in much more subtle ways than it is usually credited. It is not quite the brute it is thought to be. What's more, *Hox* genes might not be very different in kind from all the other regulatory genes.

I have already mentioned several multifunctional roles that *Ubx* can take up in *Drosophila* development. I have also stressed the 'illogical' network of interactions between all the genes involved in early development. Most of these genes produce regulatory proteins, so they all influence each other's activities in a bizarre network of arrangements. I drew you a diagram, Charles, in an earlier correspondence whose sole point was to illustrate just some of the comings and goings that take place from the maternal *bicoid* gene right through to the *Hox* genes. If we throw in what is known about target genes, some of which also produce regulatory proteins or signalling proteins, then Akam's argument that there might not be anything special or 'masterful' about *Hox* genes is well made.

Akam stresses that the compound promoter of *Ubx* means that it can get involved in a wide range of activities, some of which will have smaller and more subtle effects than its role in homeosis. Indeed, to everyone's surprise, it turns out that the promoter of *Ubx* is a massive 100,000 nucleotides of DNA. Different subregions of the promoter with different types and numbers of binding sites are responsible for the expression of *Ubx* in a given time and place during development. For example, at least seven different subregions, each typically having between four and six binding sites, initiate *Ubx* activity in the very early stages of embryo development, after the relevant regulatory proteins have sat on their binding sites.

Through the use of compound promoters, *Ubx* in *Drosophila* can be seen to be expressed in fine-grained patterns within segments giving rise to segment-specific phenotypes. So Akam argues, correctly, that the number of different segment phenotypes is limited not by the number of *Hox* genes, or by the number of their permutations, but by the number of different subregions in the compound promoters of *Hox* genes. The full range of complexity is not known because the role of *Ubx* in other developmental processes, such as the development of the nervous system and other internal soft tissues, has not been fully explained. One proposal is that there may be up to a hundred different subregions responsible for the whole gamut of *Ubx* involvement in development. These subregions would involve thousands of binding sites for a range of regulatory proteins.

Given this background, and given what we know about the binding sites in other promoters, we can be confident that the *Ubx* binding sites will be both shared between subregions and shared by the promoters of unrelated genes. In this way, the modular, redundant and autonomous nature of binding sites reflects the activities of our usual turnover mechanisms. Indeed, some *Hox* mutations are known to result from the frequent gain and loss of genetic material in promoters because of the turnover mechanism of transposition.

It is my contention, as I shall show soon with my proposal for the evolution of centipedes, that the turnover events involved in the continual restructuring of promoters need to be taken on board if we want a comprehensive understanding of evolution.

Centipedes, snakes and butterflies

Most of what I'm about to say will concern centipedes, as we've already begun that discussion. But in principle, I could hinge my remarks on other 'bizarre' systems, such as segmented and legless snake bodies, or butterfly wing patterns. Indeed, what I want to propose could, in principle, apply to the evolution of any other biological system. I will occasionally be referring to snakes and butterflies as additional examples, so let me spend a few words on these systems.

Snakes are recognized the world over as having a very distinct body shape that I do not need to spell out. Pythons, for example, have over three hundred vertebrae in their backbones with ribs on every one, anterior to the vertebrae carrying the minuscule hindlimbs. The entire trunk of the python is an elongated thorax. I have just described to you how the backbone is divided into three main sections in species such as ourselves, the cervical, thoracic and lumbar, each with distinctive vertebral forms plus other bone structures. Species of vertebrates, such as birds and mammals, can differ in the number of vertebrae that make up a particular section of the backbone, and the length of a section depends on the extent of activity of particular *Hox* genes in the developing backbone. The transformation of the majority of snake vertebrae into those typical of the thoracic region is also due to changes in the activities of *Hox* genes.

In centipedes there has been a massive repetition of segments taking their legs with them, whereas in snakes the transformation led to an equally massive expansion, but without legs. Can both of these 'parallel' events be explained by gradual natural selection acting alone, in ways we have discussed earlier?

Let me say a brief word about butterflies before I entertain this key question. I have already described how the eyespot on the butterfly's wing is a result of the transfer of a complete regulatory circuit, normally involved with wings and legs, to a 'focus' in the middle of the wing. It is as if the concentric circles of different pigments resemble a leg that has been flattened from its end so that the smallest central ring, if pulled out, would be the farthest part of a leg.

While this sort of information is important, it does not directly explain the fifteen thousand species of butterfly and moth that have evolved over the past hundred million years, each with recognizably different colour markings on its hindwings and forewings. These colour patterns have been shown to help in camouflage, warning signals, sexual recognition signals and many other important features of butterfly life cycles, as they interact with their specific environments. In the traditional view of evolution by natural selection only, all such functional features are called adaptations. Indeed, some species can change their wing-colour patterns from season to season, which is often interpreted as behaviour adapted to varying temperatures.

A handful of genes are responsible for a very wide variety of wing-colour patterns, possibly involving all fifteen thousand species. The relationship between the few genes and the many effects will no doubt reflect the usual story, which I have been emphasizing throughout, of novel permutations and subtle qualitative interactions between genes following on from TRAM features, as they control one another's activities. Moving from one pattern to another, or from one segment structure to another in snakes and centipedes is not as difficult, cumbersome and time-consuming as might be supposed.

Centipedes and lavatory rolls

The first thing we need to decide about in centipede evolution is: what is the unit of phenotype that is evolving? Is it the whole segment complete with legs, or is it some small subsection of a segment? As I have explained, *Hox* genes are capable of doing both jobs: that is, of affecting whole-segment identity (the homeotic transformation) and of being involved in more detailed patterns of cell identity below the level of the segment. In keeping with a generalized view of evolution by natural selection acting alone, it has been argued that the small, more subtle differences are more likely to be the basis for long-term evolutionary change than the 'hopeless' gross homeotic transformations.

But it is harder to explain the centipede's legs if we begin with the smaller *Hox*-influenced effects below the level of a segment. The repetition of segments in centipedes or snakes is more likely to have involved a whole segment than smaller slices of a segment. It is reasonable to assume, given that the segments of centipedes or of snakes are identical, that there was some initial restructuring of a segment by *Hox* and other genes, but that once this had been achieved, the new segment was repeated as an intact module. We can call this the 'lavatory roll model' of evolution, and it could be used to conceptualize many other evolutionary features besides centipedes and snakes.

Before I enter into the evolutionary significance of this, I have to admit that it is not known why a 'lavatory roll' process of segment proliferation took off in centipedes and snakes. But the key question from an evolutionary perspective is why such a proliferation of segments spreads through a population. It is my contention that, whether or not the spread was by natural selection, molecular drive or neutral drift, or some combination of them, the increase in number of segments, at any given time, would have needed to be incremental. No centipede suddenly acquired 173 legged segments. No python suddenly expanded to 300 thoracic segments.

The problem for natural selection is not really one of the size of the phenotypic change, but of the adaptive value of the phenotypic change as segments increase up to 173, in some species. You tried to solve this problem by supposing that intermediate levels are 'preadaptations' performing unknown functions that were, at each and every intermediate step, of adaptive value. That is to say, each intermediate step, say one segment at a time, spread in a population because it improved the relative reproductive success of its bearers. This amounts to saying, and I hope you don't mind me being so blunt about this, that what survives, survives and, given that we only have natural selection at our disposal, what survived must have had new structures or behaviours that improved their reproductive success.

Adoptation: a missing term in the science of form

There is an alternative way of thinking about the evolution of new structures and behaviours in which the mechanism of spread is initially caused by molecular drive and not natural selection. I therefore propose that we use a different word for the molecularly driven new functions. We cannot use the word 'adaptation' because it is now inextricably bound up with natural selection. We cannot use the term 'exaptation' because it is bound up with neutral drift, along with other processes, and means a later acquisition of a new function, often from an old function.

I propose to use the word 'adoptation', meaning a 'choosing' of a new component of the environment, by a slowly changing population of individuals. This adoptation is itself intimately bound up with how the genetic cohesion of a population is maintained during molecular drive. I have gone into this in quite some detail in a previous letter, so I hope you haven't lost it or used it for pressing plants.

A brief memo

For molecular drive to work, there needs to be non-Mendelian turnover among redundant and modular genetic elements. Turnover is an umbrella term reflecting the gains and losses of genetic units by transposition, gene conversion and unequal crossing-over, among others. The continual activities of turnover can lead, in the fullness of time, to the spread of variant genetic units, or an increase or decrease in the number of genetic units, through a sexual population. This is because the turnover mechanisms invariably operate between chromosomes.

As I've explained in detail, if gene conversion occurs between genes located on a pair of chromosomes, each gene acquires the same sequence after starting off with different sequences. After sex, each member of the pair of chromosomes enters a different individual at the next generation, where it meets up with a different chromosome contributed from the other parent. Gene conversion can then occur again in the new individual. Whether gene conversion is biased to occur always in one direction, or unbiased as to the direction of conversion, it is possible for the combined effects of gene conversion and sex to spread a new genetic structure through a population with the passing of the generations. This process, which I've called molecular drive, is operationally distinct from natural selection and neutral drift because it results from internal turnover in the genome and not from external turnover out in the environment. Please refer to my earlier letters for full details.

The vital thing about these turnover events is that they don't just contribute to the range of promoter configurations that selection can play with. In terms of your two-step process of evolution, which we have discussed at length, the non-Mendelian mechanisms can both create new promoter configurations and assist their spread through a population. They do not simply produce one-off mutational variety because their progression through time would then depend solely on their selective effects on reproductive success or on the vagaries of neutral drift. In the case of reproductive success, they would give rise to new functions called adaptations; in the case of neutral drift, initially at least, their effects on function would be neutral.

The promoters of genes, with their multiple and modular binding sites for a variety of regulatory proteins, are ideal breeding grounds for turnover. In the several cases where promoter differences in a given gene have been examined between species, it is clear that restructuring has involved changes in the numbers, sequences, positionings and orientations of binding sites. What's more, and importantly, binding sites to a given regulatory protein are shared by unrelated genes. All these features indicate a variety of turnover mechanisms moving binding sites from one genome position to another.

From what we know of some promoters in *Drosophila*, mammals and sea-urchins, the turnover mechanisms most likely to be involved in their generation and spread are slippage, unequal crossing-over and gene conversion. Transposition is the most important mechanism for distributing binding sites to the promoters of unrelated genes.

Is a reconstructed promoter likely to have detrimental effects on the 'regular' functions of the organism? My answer to this question rests on modularity and redundancy. The point of modularity is that a part of the phenotype can change, quite dramatically in the case of homeotic transformations, without any overt collapse in the rest of the individual. Similarly, subsections of compound promoters are autonomous and short range in their effects. In other words, incoming subsections carrying binding sites to regulatory proteins not normally found in the promoter of a given gene can produce an additional expression pattern of the gene, without disturbing the 'regular' functions of the promoter. Compound promoters are indeed composed of autonomously acting, once mobile, subsections that give a great deal of flexibility to the control and subsequent activities of the relevant adjacent gene.

Modularity is the key to understanding the effects of turnover on evolution. The complementary phenomenon is redundancy. I have already emphasized the importance of redundancy in earlier correspondence, when I described the effects of internal buffering by redundant genetic systems. I discussed buffered systems to provide a plausible case for molecular coevolution. As you might recall, that case envisaged an interaction between natural selection and molecular drive. I likened this to the trick of changing the functions of an aeroplane while it is flying in the air. But I want to leave out molecular coevolution in this particular account of centipedes' legs and the process of adoptation.

Back to centipedes

It is therefore possible to envisage a restructuring of a given *Hox* gene's promoter so that one or more segments are added in the development of our proto-centipede. When this restructuring is due to the activities of turnover mechanisms, as it invariably seems to be, then the extra segment can spread in a small number of individuals via molecular drive. These individuals can be expected to be related because they will have descended from the parental genomes in which the restructured promoter first arose. The continual activities of turnover would ensure that the restructuring was not a one-off event, restricted to one individual.

So far, we have got a newly reconstructed promoter spread to a few related individuals. To avoid any arguments about how many segments such a new *Hox* promoter gives rise to, I want to suggest that it is one or a few at most, but not 173 segments. My reason for this has less to do with whether natural selection could deal with grossly formed 'hopeless monsters' than with the crucial need for parental genomes to give out the same development 'messages' to a newly fertilized egg.

Singing from the same hymn sheet

I have told you, Charles, of my proposition that one of the key effects of sex is to ensure that the 'messages' emanating from the two parental genomes are compatible. The 'cohesive' genetics of molecular drive combined with the mixing consequences of sex at the population level ensures that, in general, everyone is singing from the same hymn sheet.

In other words, if we have, say, some half-way stage in the progression from a few to 173 centipede segments, no one is left in the population with only a few segments and no one has raced ahead to 173 segments. The combined efforts of sex and molecular drive, even without the involvement of selection in the establishment of molecular coevolution, ensures that a newly developing centipede embryo is being told by the two parental sets of developmental genes to make a roughly similar number of extra segments. There should be little ambiguity or confusion in the messages. Nor should there be ambiguity or confusion at any stage in the push to 173.

Naturally, I don't know why the number of segments should stop at 173 in some centipede species and at other lesser numbers in other species. It might be the result of a cessation of turnover or excessive strains in the whole developmental system, about which natural selection would no doubt have its say. If we didn't know that pythons with 300 thoracic vertebrae existed, would we ever have believed that a 300-segment 'monster' was possible?

I want to emphasize, Charles, that I'm not saying 'anything goes'. New developmental systems are acceptable only if they are compatible with the developmental events that make up any given species. But it might not matter too much what precise developmental path a species takes. As I've said before, we know that there are over five thousand different ways of making *Drosophila* species. If turnover events among their TRAM promoters had been slightly different, at different periods of their evolution, then there might have been five thousand alternative forms of *Drosophila*, most of which might have been equally successful as the five thousand that did evolve and became established. If we throw into the pot the involvement of selection in molecular coevolution and the facility for adoptation (see my next comments), we begin to view the evolution of life as a more relaxed, versatile and tolerable process.

In my view, therefore, evolutionary success can be measured according to internal developmental factors and not only according to external environmental factors in a world governed solely by natural selection. This leads to my final argument that there could be a much looser relationship between organisms and their environment than is generally supposed.

Choosing the environment

If I am right that the success of newly elongated centipedes, to take just one example, is measured in terms of developmental compatibilities between parental genomes in sexual populations, and not only through the provision of 'solutions' to environmental 'problems', then it could be that the elongated centipedes have a much looser relationship with their existing environments than we have previously been willing to entertain.

If we accept this possibility, we can proceed to the next step. This is to suggest that the very 'looseness' of this relationship means that a population of elongated centipedes can *adopt* some part of the environment that was previously inaccessible. In other words, the similarly constructed new population of centipedes might be able to crawl into places that were previously barred, or to escape predators that were previously troublesome. As with your process of naturally selected adaptations, Charles, there is no end to the adoptational possibilities.

We could provide adoptational explanations for the fifteen thousand different wing patterns on our fifteen thousand different species of butterflies and moths. I believe that the changes by turnover in the reconstruction of promoters of the relevant genes involved with 'eyespots' and their subsequent spread might have led to the establishment of fifteen thousand alternative patterns, each of which might have been subsequently exploited in sexual signalling, warning patterns and so on – just like the fifteen thousand patterns now considered to arise through natural selection.

Adoptational landscapes

One of the classic metaphors of evolution is the 'adaptive landscape', as first proposed by Sewall Wright, America's foremost evolutionary geneticist. This is a mountainous terrain where each mountain represents an 'adaptive peak' on which a species sits. Some species are on lower peaks. This means that opportunistic selection has taken them by the hand up the nearest slope. But their limited genetic repertoire means that such species get stuck on lower peaks. Any further mutational changes are of little use because to get to a higher peak of adaptation the population first has to descend to less adaptive valleys and this isn't allowed in evolution governed solely by selection. So each species on its peak is by no means perfect, but it has done its best.

The image of the adaptive landscape is quite like the lock-and-key imagery that I used earlier. Under molecular drive the landscape metaphor becomes superfluous. Species are not trapped on peaks. Indeed, cohesively evolving populations can seek out and choose suitable peaks. A better image might be the heaving surface of the ocean caused by the inner turbulence of the genes matching the outer turbulence of the environment. There is constant change in dynamics because populations are neither locked into a niche nor necessarily stuck up a fixed adaptive peak.

If adoption leads, like adaptation, to functional relationships between organisms and their environment, it is no longer possible simply to look at organisms in their natural habitat and decide in advance how much adaptation and how much adoption might have taken place. It is my belief that adaptation, adoption and exaptation will all have been involved in the relationship between a given organism and its 'niche'. But let's keep them separate for the moment to clarify the distinction.

Under adoption the organism, in a sense, chooses its own niche, while in adaptation the niche chooses the organism. There is a fundamental distinction here. The idea of 'choosing' is not new and has been rightly stressed by evolutionary biologists such as Richard Lewontin of Harvard and Patrick Bateson of Cambridge in their emphasis of the active dynamics between organisms and the environment.

Keeping left!

Perhaps I can best explain this, Charles, in terms of driving on the left or on the right-hand side of the road. We no longer move in horse-drawn carriages in our century, but in fast-moving cars in which it is essential to drive consistently on one side of the road. Failure to obey the rule is disastrous. Maybe you had the same rule in your day. For some historical reason, driving on the left took precedence over driving on the right in the United Kingdom whereas in countries like France the reverse happened.

Let's think about this situation. The two systems, in themselves, are equally good as long as there is a uniform cohesion among all drivers in a given country. I doubt very much if such a cohesion, whichever way it turned out to be, resulted from a long process of trial and error involving the frequent selective elimination of one set of drivers. Instead, the left or right driving rule was an historically based decision. As with all my previous analogies, I do not want to push this too hard. But for the sake of completeness, let's imagine that we play God and turn around a particular car to drive in the 'wrong' direction, while all the others are driving in the 'right' direction. Inevitably, there would be a dramatic loss of reproductive fitness of the driver as the car crashed into the oncoming traffic. Now, if this were an experiment in biology, such a loss would be taken as proof of natural selection. But natural selection was not involved in the original establishment of the left or right driving rule, even though a reversal of direction suggests that this is how the rule got established.

There have indeed been experiments in which investigators have shortened or lengthened the tails of a species of bird to measure the reproductive success of individual males, in which tail length is considered attractive to females. It turns out that long tails are better than short tails in this respect. But this does not mean that tail length arose solely by natural or sexual selection.

Let's return to the adoptational possibilities of evolving centipedes and snakes. Extra segments are not unusual or unexpected given what we currently understand about genes such as *Hox*, or regulatory genes in general, armed with their compound promoters. Clearly, body segments can increase in number, as modular units, as long as one parental genome does not make 173 segments all in one go. If such gross incompatibility is avoided, as it should be given the rates of turnover, then gradual accretions to body length could lead to new 'adoptations'. But this will not be the end of the story because I'm sure that molecular coevolution and adaptation would also have occurred, giving an intriguing mix of forces involved in new relationships between the organisms and their environment.

I do not believe that centipede and snake trunks are different in kind from the modular construction of almost everything else in biology, including human hearts and brains, under the influence of regulatory networks with TRAM characteristics. It might seem more difficult to explain the brain or the giraffe, simply because there are so many parts involved. But as long as these parts are modular, the same processes can occur. Brains and giraffes are as much like giant

Lego sets as segmented trunks are, and the same arguments can apply. At the very least, the known modularly constructed back-bones and hearts of giraffes, under the control of *Hox* and other genes, would get us off to a good start. Maybe the giraffe, after all this time, is an adoption when a taller set of individuals exploited their molecularly driven extra height to munch leaves at the tops of trees, or for the males to fight more successfully in their sexual displays. Just a story!

I will finish on this note because I'm sure you will have plenty to say about adoption. It is only one letter different from adaptation, so it isn't too dissimilar an outcome, even if the underlying forces of molecular drive and natural selection operate in fundamentally different ways.

I've also set aside for the moment your questions about genetic determinism and free will. This is a very important topic and I'm glad you've raised it. I do, indeed, have several things to say about this very contemporary debate.

For the moment, however, I will sign off.

Gabby

Westminster Abbey

Dear Gabby

Your 'adoptation' letter has finally bridged what was becoming a worrying gap in my understanding of your proposals about the nature and evolution of biological functions. I do have some reservations about adoptation that I will come to shortly. But before I start, I want to remind you that you haven't yet answered my concerns about genetic determinism and the opportunistic nature of natural selection. I do not have any particular axe to grind about determinism, but I would like you to be clear about your views because there seems to be an ambiguity over this.

The issue of 'opportunism' too needs to be resolved. As I have said earlier, my theory of natural selection does not, or should not, suggest that anything in biology necessarily makes sense or appears 'logical' in its construction. I can see where your irritation might creep in when, for example, eyes are taken as perfections of biological engineering that could not have been constructed in any other way than by natural selection. This view gives the impression that out there in the big wide world there is always enough variation for selection to be able to pick and choose what it needs, like a child in a sweet shop. Natural selection is as much restricted by the available variation and constrained by historical precedence as your process of molecular drive, if I've understood that completely. Both systems give rise to novelties that make the best of the available raw materials. If that means promoting a new structure that happens to function in some highly 'messy' and 'illogical' manner, not befitting any self-respecting engineer, then so be it. Getting stuck on a low peak in the adaptive landscape is symptomatic of this situation.

Now about 'adoptation', first let me say that the word is well chosen. It nicely conveys the act of 'choosing' new niches by a group of slowly changing organisms, as they actively participate and exploit the world around them. And who can blame them for that?

My main concern might sound a bit philosophically pedantic but hopefully you will be able to clarify your point if I approach the problem as follows.

Is there any real distinction between adaptation, adoptation and exaptation in how new relationships between organisms and their environment are set up? Let us look at natural selection. We have both stressed the point that mutations are random and undirected towards any future adaptive needs. There is no justification for bringing in Lamarck through the back door. The deeper significance of this is that the resulting individuals, even if they are accumulating by natural selection, can be said to be 'adopting' their environment. The initial variant is produced, willy-nilly, but its propagation depends on there being some new relationship between it and the local environment that improves its relative reproductive success. There is a lucky match between the new variant and some new component of the environment, and this could be described as much as an adoptation as an adaptation.

From what you tell me about molecularly driven adoptations, the

same basic luck in organism–environment matching is also occur-
ring there. You describe this matching as the organism choosing
some previously inaccessible part of the environment. But I am sug-
gesting that it is no different in kind from the matching required of
natural selection.

I want to make the same argument about exaptations and then to
hear back from you. From what you have told me, an exaptation
could arise from neutral drift, although I realize that there are other
ways in which a new structure can be spread in a population whose
later functional co-option fits the definition of exaptation.

An amusing example that has just occurred to me, as I lie on my
back looking up at the ceiling of the Abbey, is the many uses to
which the architectural spandrels have been put. I'm referring here
not to the structures required to keep the building up or the place
dry, but to all those inevitable 'gaps' in the building where the more
essential structures come into contact. Many of these gaps have
been filled in with decorative features that are not absolutely neces-
sary for the building to remain standing. This gap filling, taking
place after the building itself is established, produces embellish-
ments and ornamentations that could be called exaptations, as
opposed to the purposely built adaptive structures. I've been reading
so many of the papers that you've been sending to me that I'm no
longer sure which ideas, including this one, are originally mine – not
that it should matter in the long run.

But let me stay with neutral drift as a way of achieving an exap-
tation. As the word 'neutral' implies, a new variant structure or
behaviour could spread through accidents of sampling, unimpeded
or unaided by selection, because it is neutral with respect to repro-
ductive success. But once it has spread over the generations, a
fortuitous change in circumstances might switch the variant struc-
ture or behaviour from being neutral to being functionally required
for reproductive success. What I'm trying to stress here is that there
is the same fortuity in match between an evolved new organism and
some environmental variable built into the process leading to an
exaptation. The genetic variants, as they appear in individual organ-
isms, are no more preordained to match a given environment under
exaptation than they are under natural selection or molecular drive.

Looked at in this way, all three mechanisms of evolutionary change seem to be in the same boat. Mechanistically, they differ widely in how they 'promote' the spread of novelties. But the final functional products could be said to be the same: a new, fortuitously arising relationship between a new organism and a new niche. If we accept this, we don't need to use the separate words 'adaptation', 'adoption' and 'exaptation' for one and the same outcome.

As I said, Gabriel, I realize that this is a bit philosophical and perhaps a little removed from the pragmatism that always guided my own biological thinking, and probably your own. But I might have provided some useful hooks on which you can hang some further clarifications.

Your most sincere friend

Chas. Darwin

THE UNKNOWABILITY OF DNA

Dear Charles

I think we are rapidly coming full circle in our correspondence. I've thoroughly enjoyed relaying to you what's happened in the brave new world of biology and genetics, and letting you have my opinions of what it might mean from an evolutionary point of view. I have to say, though, that I have by no means covered everything of interest – I'm not a walking encyclopaedia! I can gather from your responses that you too have enjoyed this exchange of ideas a hundred years and more after your internment. Maybe we could all do with an enforced long rest in some peaceful abode, to emerge with new looks at old problems. If you are not too averse to the idea, and if I can pluck up enough courage, maybe the public at large will enjoy reading our exchanges if published.

I agree with your view that adaptation, adoption and exaptation have something in common that could make separate words for them unnecessary. But finding a common word to embrace all three would be difficult, given the historical precedence of adaptation and its close association with natural selection. I have always considered it important to distinguish their differences in order to explain three different forces involved in the inception and establishment of biological novelties. It is rather crucial for the pragmatic approach to biology that I'm trying to emphasize, that we do not succumb to the temptation, for example, to look at a giraffe, happily living and reproducing in its niche, and conclude, hastily and superficially, that the giraffe is nothing more and nothing less than an adaptation having arisen solely through natural selection.

That sort of conclusion was par for the course when all we had was natural selection. But from what we currently understand about the relationships between genes and phenotype, about the Mendelian and non-Mendelian mechanisms of genetic transmission and turnover, and about patchy and fluctuating populations in their natural environments, we can expect giraffes to be a mixture of natural selection, molecular drive and neutral drift, with their potential functional products, definable as adaptations, adoptations and exaptations. The business of contemporary biology is to quantify the relative contributions of each of these three forces to the evolution of any given novelty or new species. This will not be easy – indeed, it could be said to be an exercise in futility given what we already know about the potential for all three evolutionary forces to interact. Molecular coevolution, for example, can only be understood, as I've explained, as the result of a continuous dynamic between molecular drive among redundant genetic elements and natural selection on other genes that interact with the multiple elements.

In my more pessimistic moods, however, I do believe that we are a long way off investigating interactive processes. The more immediate and important task is to get biologists to recognize that there are, indeed, more complex processes underlying the evolution of what they would previously and all too easily have accepted as naturally selected adaptations.

We face an enormous problem. It is not just that professional biologists have drunk a little too readily on the teat of natural selection. The idea that all of biology can be explained by past events of differential reproduction pervades most non-biological disciplines too.

Three into one won't go

I have given many of the reasons why we must develop a broader perspective on the complex interactive forces governing biological evolution. I believe that the point can be made clearer with a choice use of words specifically designed to discriminate our three great forces, natural selection, neutral drift and molecular drive. This sentiment brings me straight to your point about the deep similarities between their respective products, adaptations, exaptations and adoptations.

If we consider the very clear distinctions between the detailed operations underlying these terms, then they cannot, and should not, be confused. You are absolutely right to reaffirm the 'random' nature of mutation as it feeds into adaptation through natural selection. There is a sense in which the fortuitous match between a randomly produced mutation and the environment can be said to be a sort of choosing by the variant individual. Perhaps it is no different in kind from the 'choosing' I emphasize in molecular drive, which takes a genetically cohesive, yet changing, population of individuals into a new relationship with the environment.

But as you said, this is all getting a bit philosophical and is in danger of distracting us from the more important operational distinctions between these two modes of evolution. The adaptational product of natural selection becomes established over time because in each generation some genetically endowed subsection of a population reproduces more successfully. The individuals concerned are more in tune with their environment in terms of their ability to reproduce successfully. Because the genes help to construct these individuals, and because genes are passed on to the next generation, the genetic basis of the adaptation slowly accumulates over time. It is easier to visualize adaptations arising through particular organisms being chosen as local solutions to problems imposed by a demanding environment than the reverse. In earlier letters I characterized adaptation as a process in which organisms are dragged kicking and screaming to particular environmental niches as life gets harsh. This might be an overdramatization, but the point is made.

It is my contention that the adoptational result of molecular drive represents a different mode of evolution. Molecular drive too relies on the 'random' production of mutations. But they are spread in a population not through the differential reproductive success of specifically adapted individuals, but through non-Mendelian mechanisms of turnover, coupled to sex. As I've described in detail, all or most of a population is slowly moved, collectively and cohesively, from one genetic make-up to another. During this shift, the changed organisms engage in a new functional relationship with their existing environment: that is, they collectively and actively adopt rather than being selectively adapted one after the other.

Let's assume that molecular drive alone had been responsible for

giraffes. We can suggest that, as the necks got longer, the whole population of changing individuals could start exploiting, in a fresh way,
the existing environment of tall trees. Given the fundamentally
different operations of selection and molecular drive by which
giraffes might have got their long necks, I don't think that the functional outcome (tall giraffes eating leaves on tall trees) can be given
the same word. The adoptation potential of a cohesively evolving
population, driven from within the genome, is different in
principle from the adaptational component underlying differential
reproduction of individuals.

Naturally, it is worth emphasizing that, if there were no tall trees
to be eaten, our molecularly driven population of long-necked
giraffes might go extinct. This would be no different from the extinction of variant individuals that show poorer reproductive success
under natural selection.

It is important to recognize that from the perspective of any given
species the environment is largely unexploited. To be a little dramatic about this, let me say that the air is there for us to exploit if
only we developed wings. The darkness of night is there to exploit if
only we could see in the dark. We do not know enough about the
environment to be able to dismiss the potential of an adoptational
'choosing' process arising among a molecularly driven population of
individuals. If we add molecular coevolution, in which selection is
intimately involved in maintaining functional standards while a
change is taking place across a population, then the final adoptational product might not be as difficult to achieve as we might
imagine.

Assuming that the evolution of new functions can arise only by
means of natural selection, it is possible to imagine an environment
as an adaptive landscape or as a tight system of locks and keys as
organisms are adapted to ever-more refined niches. Your own
metaphor was of a log full of wedges in which it would be difficult to
place yet another wedge without displacement. The wedge metaphor
is perfectly acceptable and understandable in terms of natural selection working alone. Today, however, we know that additional forces
are at play in evolution that emphasize a more tolerant relationship
between organisms and their environment and between interacting
molecules within organisms.

Similarly, the exaptational potential of neutral drift is operationally distinct from natural selection and molecular drive. I believe that Stephen Gould and Elizabeth Vrba were correct in proposing exaptation as a missing term in evolutionary biology. I want to make a distinction between adoptation and exaptation here. Again it is about mechanistic differences. An exaptation can be viewed as the acquisition of a new and useful function once the novelty has spread and once the environment has changed. Adoptation, however, can begin during the earliest stages of a molecularly driven homogenization and spread of a new structure through a population. If we take our giraffes as examples, a gradual change in neck length is potentially exploitable as an adoptation from the very beginning because the newly endowed population of longer-necked giraffes are able to reach slightly higher leaves. And so on through all the intermediate stages from a short-necked population to a very tall-necked population. It would therefore be confusing to describe the functional outcome of molecular drive either as an exaptation (because it did not spread initially as a product without a function) or as an adaptation (because it did not spread initially as a result of differential reproduction).

A psychological difficulty

I believe that the understanding and acceptance of new ideas is, like evolution itself, constrained by and contingent on what we have already established in our own minds. Imagine, dear Charles, that you had been brought up on nothing but molecular drive. All professional courses in evolutionary genetics and all popular accounts of evolution (giraffes, beetles, intelligence, sexuality and so on) had molecular drive at their centre. You had been immersed in the antics of the non-Mendelian mechanisms of turnover, in all their fascinating and bizarre details. You understood that these gave rise to cohesively evolving, sexually reproducing populations, so that tolerance and cooperation were the order of the day between molecules within individuals and between individuals and their environment. Adoptation was the order of the day as an explanation for the complementary relationship between an organism and its niche. And now imagine that all of this had been common knowledge for 150 years.

Then, having learned all of the above over a lifetime's thinking, you were confronted by my exhortation to accommodate Mendelian mechanisms of inheritance, random processes of meiosis and the genetics of the Hardy–Weinberg equilibrium, Malthusian pressures of overpopulation, differential survival and natural selection, and the novel feature of adaptations. You might very well find it hard to take both intellectually and emotionally at one sitting. Familiarity does help.

So if you are wrestling with difficulties (and I don't mean this in a patronizing way), you will surely recover as you begin to think through and apply alternative perspectives to whatever biological problem you have in hand. My main objective is to sow a seed of doubt in your mind about the origin of biological diversity and the nature of nature.

Darwin's finches: what we know and what we don't know

Let me give you, Charles, an example that might bring this message home more readily. The story revolves around the many different beak shapes of thirteen species of finch that you observed in the Galapagos Islands. This story is so well known that they have been called 'Darwin's finches' and our textbooks carry the obligatory picture of a series of finch faces, all with beaks of different shapes and sizes. Each shape and size nicely correlates with using the beak to obtain food from different sources. Big, strong beaks are useful for breaking up large nuts, and thin, smaller beaks are useful for probing into fruits and the like. For each beak there is an exploitable source of food. What's more, different islands are occupied by one or a few of the finch species. I don't have a diagram of your finches at hand, but the enclosed diagram of the Hawaiian honeycreepers shows a similar diversity in beaks.

Although at first you failed to realize the significance of these observations, you did eventually propose that each beak shape had been naturally selected on a given island to exploit the specific sources of food available. This is a perfectly adequate explanation and is generally assumed to be true to this day. Indeed, recent detailed observations and experiments have shown dramatic swings in reproductive success if the relationship between beaks and food sources is naturally disturbed by seasonal shortages.

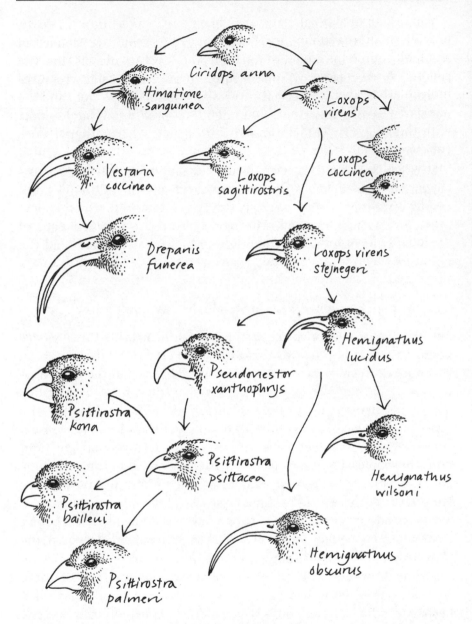

Hawaiian honeycreepers *This is just part of a large number of related honeycreepers. Similar to your finches on the Galapagos Islands, each species is distinguishable by the shape of its beak, and these beak shapes vary enormously. Each shape is useful for obtaining food from a different source and is considered to be an adaptation produced solely by natural selection. Could they be exaptations or adoptations? Or a mixture of all three processes of evolutionary change: natural selection, neutral drift and molecular drive? This picture is modified from W.J. Bock,* Evolution **24**, *704–722; 1970.*

You are probably already anticipating, Charles, how I might provide an alternative interpretation of such matches between beaks and food sources as adoptions. I do not want to spell out the conceptual details again, except to say that some of the assumed molecularly driven changes in beak shape and size (which I'm sure one day will be shown to be under the influence of regulatory genes with their TRAM characteristics) might produce beaks that allow a cohesively evolving population to exploit an existing food source that was always available. And there could be many occasions when changes in beak shape and size did not lead to adoptions and could not be tolerated, either through molecular coevolution or in any other way. In these cases, the unfortunate populations can be expected to have gone extinct.

Our problem is that we only see the winners in the evolutionary game, and this applies to all three forces of evolution: natural selection, molecular drive and neutral drift. We could conclude that all of evolution is the survival of the luckiest. Maybe this is what you meant when you suggested that there is a common feature between adaptation, adoption and exaptation that would require one, as yet unknown, word (see my diagram).

An operational tall order

Finally, and importantly, if we really wanted to know in depth the evolutionary origins of the finches' beaks, we would need to know the genetic basis of beak development. How many Mendelian and how many non-Mendelian genetic units are involved? What turnover mechanisms are involved? What are their rates, biases and units of operation? How much molecular coevolution might have occurred? How much neutral drift can be seen in the genetic sequences? What are the precise ecological dynamics affecting the survival and reproduction of finches? What key factors in finch life cycles and development affect such dynamics? This is a tall order for finches and for any other biological feature in any other organism. Clearly it requires data gathered from a variety of levels from the molecular to the ecological. The same full set of procedures would be required if we were to dissect the evolutionary forces involved in our human brain functions, such as consciousness and language.

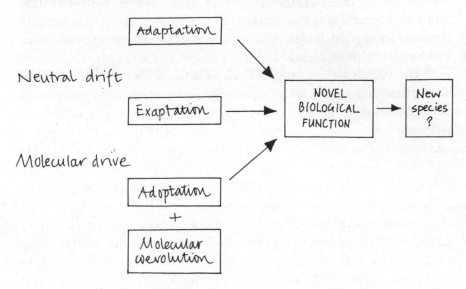

Three processes; three outcomes; one product *A novel biological function can arise by any one of three operationally distinct processes: natural selection, neutral drift and molecular drive. Each process has three distinct outcomes: adaptation, exaptation and adoptation. Additionally, the response of natural selection to molecular drive leads to molecular coevolution. In all probability, the internal pressures generated by flux in the genetic material, coupled to the external processes generated by flux in the environment, produce evolutionary novelties and potential new species that are a mix of all three processes.*

What is no longer permissible, in the context of our new understanding of the nature of the behaviour of genetic material, is to measure the contribution of any given trait to reproductive success and then to claim that the trait arose via natural selection. The adoptational and exaptational features of biological functions could be just as much involved in current survival and reproduction as adaptations. As I explained in my left- and right-hand driving rules, no validation of a process of natural selection could be extracted if we turned a car around to face the 'wrong' direction and watched a dramatic drop in reproductive success. Evidence for natural selection can be garnered only from detailed examination of the appropriate genetic systems underlying any given trait. For example, there is considerable evidence for the participation of selection in the establishment of molecular coevolution.

These are exciting times for evolutionary biologists. The days have gone when we could restrict our views about the evolution of new forms and behaviours to natural selection alone. Molecular drive is involved at several levels, forging a link between genetic turnover and functional novelties.

At the very least, as Samuel Beckett said in his novel *First Love*: 'that other links exist, on other levels, between these two affairs, is not impossible'.

As ever

THE EVOLUTION
OF INDIVIDUALITY

Dear Charles

I'm shortly to go on a summer holiday in the French Pyrenees and, rather than wait for your reply to my previous letter, I've decided to send you another letter and post the two together. If nothing else, this saves on postage! This second letter summarizes many of the opinions I've been expressing in our correspondence on issues of genetic determinism and the supposed dilemma of free will in humans. So, in a sense, you can relax because I'm not going to throw in any more contemporary biology. I realize that there are large areas of sophisticated biology that I have not covered, but I hope I've erected enough of a factual framework on which to hang some alternative conceptual approaches. Food for thought, that is.

Determined to death

You are right, Charles, to point out some lingering ambiguity in my arguments about the genetic basis of individual development and my strongly held stance that the genes cannot be the eternalized units of selection. I recognize that there is a potential conflict here – I cannot write reams of descriptions about the way genes contribute to individual development and behaviour and then opt for a philosophy of biology that does not have genetic determinism written all over it.

I think there is, indeed, a widespread confusion in my day between the role of genes in evolution and the role of genes in development. I have attempted to show, throughout my letters, that there is no one-to-one relationship between particular genes and particular bits of an individual organism. This is because the majority of genes can participate in a very wide variety of functions and each function is the result of a wide variety of genes. During development a many-on-many dynamic emerges from the intricate webs of interaction between promiscuous, modular and redundant genetic elements. I cannot summarize it otherwise than repeat my little aphorism that in biology $1 + 1 = 7$. We don't have a theory of interactions and until we do we cannot have a theory of development or a theory of evolution.

I am not saying that the phenotype is beyond the reach of genetic interactions. There is nothing mysterious about a given individual. It is capable of being described theoretically from conception to death in terms of the full range of genetic and environmental influences.

Such a description, however, would need to be done after the individual has emerged, not before. In other words, we could not predict any given outcome even if we had all the inherited genes of an individual strung up and rung out to dry in front of us. We cannot determine in advance the future development of a newly fertilized egg. There is an unknowability and unpredictability about DNA, as an inherited molecule, that undermines any notions of individual genetic determinism.

Before I expand on this, I want to probe the concept of the 'individual as nothing but chemistry' a little further to show what a trivial and essentially uninteresting argument it is. It has been put forward as a serious line of thought leading to the proposition that, as individual humans, we have little free will within broad categories of behaviour. Edward O. Wilson, the father of evolutionary psychology, has made such an argument in his book *On Human Nature*.

Tossing human nature

To make his point clearly, Wilson draws an interesting comparison between the tossing of a coin and the ultimate predictability of our predetermined human nature. It is generally assumed that a tossed coin can land in one of two unpredictable alternative positions: head or tail.

Not so, says Wilson. If we could only measure the physical properties of a particular coin down to the arrangements and energies of all of its chemical elements, as well as measuring the forces of the tossing thumb, the local air currents, the force and angle of flip, the microtopology of the table, what the tosser had for breakfast and to whom he or she last spoke and in what frame of mind – then, fore-armed with the totality of this information we could predict, with a probability much higher than chance, what the likely outcome is to be.

In the same manner, Wilson claims that if we knew the personal molecular history of a particular individual from the moment of con-ception, and had a complete description of all the external influences on that individual, then we could know the unknowable. A past, present and future map of the individual, in terms of all of that indi-vidual's decisions and choices, could be constructed. To be precise, Wilson recognizes that in the central nervous system there are many 'hundreds or thousands of variables to consider, and minute degrees of imprecision in any one of them might easily be magnified'. As a result, 'it may be a law of nature that no nervous system is capable of acquiring enough knowledge to significantly predict the future of any other intelligent system in detail' and that in this sense it cannot eliminate free will.

The argument is that because the full description of the workings of our minds cannot actually be done, given the limited capacities of the minds engaged in the exercise, 'we are consequently free and responsible persons'. However, Wilson goes on to say that 'the paradox of determinism and free will appears not only resolvable in theory, it might even be reduced to an empirical problem in physics and biology'.

I find the argument trivial because it is essentially specious. Biologists are no less materialists than chemists and physicists. They deal in real molecules, alive and kicking at every turn in the massively complex processes that have led to species diversification at the evolutionary level and to the development of form and behaviour at the individual level. No serious and credible biologist is in the business of explaining human nature in terms of Scotch mist or Yorkshire fog.

Of course, *of course*, if we knew of every molecular twist and turn in the development of an individual, even including the billions of neuronal connections that make up the brain, then we would arrive at a completely knowable description of the individual, as we did of a particularly tossed coin. But there is no conceptual glory to be had in that totality of knowledge. All we have done is to decompose (literally) a given individual into its chemical constituents, including, I am willing to concede, the sequence of molecular interactions that took place in the particular development of that particular individual.

Predicting with hindsight

The exercise, as proposed by Wilson, is close to a tautology. An individual, as an intact phenotype, exercises free will in ways that are unique to that individual. This behaviour, it is claimed, can ultimately be explained by the unique unfolding of genetic and environmental interactions in that individual. So what have we gained? An individual chooses to wear a green hat one day, or write *On Human Nature* another day – choices that are circumscribed by the totality of internal and external influences on that person.

In my view, the deconstruction and then reconstruction of an individual is irrelevant to the central issue of individual predictability and free choice, whether the individual is viewed as a whole phenotype or as a phenotype in bits. The individual is the bits and the bits are the individual. Who would want to deny the circularity of this truism? All we are saying is that the individual is the individual is the individual – warts, molecules and all.

Extrapolating from nothing to nothing

Such knowledge, no matter how complete, of a given individual cannot be used to predict the form, behaviour or choices of any other uniquely formed individual. The hard-won knowledge pertains only to the specifics of the individual from which it was garnered. Should we want to predict the material basis of a different individual's form and nature, we would need to carry out the same decomposition. But by then it would be too late because the exercise is not a method of predicting how the individual might turn out. It is simply a deconstruction at the molecular level of why an individual turned out to be the unique individual he or she happens to be.

We can know what's going on *after* the individual has developed into the full glory of its uniqueness, but not what will go on *before* it has happened. We can 'predict' with hindsight, but not with foresight, even if armed with that individual's total genome of sequences. We can also make 'average predictions' about how an individual might turn out given the consensus of other similar, but not identical, phenotypes. But this too, in my view, begs the question concerning the exercise of unique choices by unique individuals. Australian chemists claim that they can break down a wine into 540 essences. But they will never be able to recreate the natural time-dependent processes of a particular grape in a particular environment.

The hunt for 'universals'

If every individual exercises free will that is influenced by its unique combination of genetical and environmental influences, and then exercises more free will over and above that, it would seem that no broad generalizations are permissible in biology about predictability. Unfortunately, however, no discipline can be pursued without its generalizations and evolutionary psychology is no exception. So the hunt is on for 'universals' in human behaviour that are claimed to be both genetically determined and predictable. As Wilson says, 'if the categories of behaviour are made broad enough, events can be predicted with confidence'. Contemporary human nature is therefore considered (within the limited perspective that evolution can proceed only by natural selection) 'as a hodge podge of special genetic adaptations to an environment largely vanished, the world of the Ice-Age hunter-gatherer'. In other words, because natural selection has acted on the behaviour of individuals who benefit themselves and their relatives, 'human nature bends us to the imperatives of selfish tribalism'.

This is a rather bleak view of human nature, from which Wilson does manage to salvage some laudable ethical code of behaviour for humans. But it is not the bleakness or the escape-hatch that worries me. There is a far greater technical problem with the issue of the 'universal' attribute. The search for 'universals' is not unique to those who wish to explain human nature by genetic determinants. Universals are also the Holy Grail for those wishing to explain human nature as a cultural phenomenon. There can be widespread commonalities in culture underpinning universal behaviour. All cultures presumably prefer food, health and order to their opposites. But the genetic argument for universals is difficult to sustain because, if genetics has ever had a consistent message, it is that genetic variation is the name of the game. If the evolutionary psychologists want to dig out the genetic determinants for this or that aspect of human nature, 'universality' is the last thing they will want to find.

There's no such thing as an average human

The only way around this problem is to resort to the idea of the 'average person', but as I shall show there is no such thing. Let's look at the genetics of blood groups. Over a dozen genes are responsible for the proteins that define our blood type. Each gene has many alternative forms (alleles) and the frequency of alleles varies enormously between populations. For example, frequencies of individual alleles at the *ABO* and *MNS* genes differ between English Caucasians, Basques and Navahos. Similarly, there is enormous variation in the genes responsible for our immune responses to pathogens, and in the repetitive arrays responsible for individual genetic fingerprints. These systems are as genetic as one can get. The only 'universal' feature is that *ABO*, *MNS*, immune response genes and 'fingerprint' arrays exist in all humans. But this is an almost meaningless generalization – it is hardly going to illuminate the genetic basis of individual variation and the predictability of free choice. When we are concerned about central issues of human nature, and the ethical and political spins that we put on them, we need to focus hard on individual human nature rather than rely on the vague truisms of the collective 'average' or even the 'universal'.

Richard Lewontin has made the telling point that, if we constructed the average genotype containing the most common alleles at the dozen blood-group genes, only one person in five hundred would possess this 'abstraction' of a genotype. If these blood-group genes determine human nature, only one person in five hundred is truly human! (Charles, this is not my ironic joke – I've taken it from the writings of the perceptive philosopher of biology, David Hull, at North Western University, Illinois.)

No one is normal

Not only is there a serious difficulty with 'universals' and 'averages' from a genetic perspective, but there is the additional problem of equating the 'average' with the 'normal'. To understand the weakness of this approach, let's just look at some of the proposed universals of 'normal' genetically determined behaviour: heterosexuality; male aggression; female social networking; stepparents beating up stepchildren; pride in ethnic group; appreciation of beauty; desire to play; curiosity and exploration; rationality; lying; laughing; feeling guilty; parental conflicts; language and so on. The more universal it can be, the more genetic it is assumed to be. But if we were really to look at human variation in many of these behaviours and define an abstract average encompassing all the most frequent variants, we might find very few individuals in existence who are endowed with the full gamut of supposed 'normality'.

The exercise as a whole is futile and is genetically misconceived. It cannot reveal in any predictive way how any particular individual is going to turn out. The individual is at the heart of society and unless we have anything useful to say about the genetic determinants of every individual behaviour (if that is what we want to do), we have no biological prescriptions for how society should organize itself.

This is a particularly severe problem when we take on board that there is no hidden 'little person' in the genetic material instructing each and every individual to adopt a universal mode of behaviour. Nor is there an alternative 'little person' hidden in the crevices of the environment. The unfolding interactions between genes and genes and between genes and the environment are unique and unpredictable for each and every individual. They are theoretically 'knowable' with hindsight, but they cannot be known in advance, even as an exercise in predicting the 'average' and the 'normal', because these do not occur in practice across the variety of human behaviours. Despite the fact that the double-helical structure of DNA has become the logo of the twentieth century, there is no hidden, preordained message in its sequences.

There's more to human nature than adaptations

There is one more fallacy about the genetic determinism of 'universal' behaviours – that the genes in us today are those that were naturally selected in the past. Arguments for genetic determinism in development and in evolution seem to go hand in hand. I don't want to labour this point, Charles, because we've been discussing other evolutionary processes leading to exaptations and adoptations at length. Our human nature, in all its manifestations and in all its individual variety, will be shown one day to be a mixture of adaptations, exaptations and adoptations, just like everything else to date that has been examined in any depth. Accordingly, some part of the historical events that make up our behaviour will be about cooperation, coevolution and tolerance, in ways that I've described.

It is against all that we are learning about the successful yet messy ways in which biology goes about its business, to continue with the belief that genetic determinism in human form and behaviour is all about grand continuities and elemental struggles between supposedly eternal, self-replicating genes.

We are individuals not groups

What I have said above is also relevant to the genetic basis of race and gender, and many other traits supposedly divisible and predictable along genetic lines. Today we can be embarrassed by the crudity of arguments that link genes to intelligence and intelligence to race or class, such as John Davenport's testimony to the American Congress in the 1920s that alcoholism, poverty and avarice are genetically determined traits of Irish, Italian and Jewish immigrants respectively. But there is an underlying invidiousness about the genetic determinants supposedly fixing our psychological make-up, sexuality and free will that is based on an outdated and quite wrongheaded idea of how genes work. This is as true when we limit our evolutionary thinking to natural selection alone as when we take a more comprehensive and pluralistic view of evolution as a product of natural selection, neutral drift and molecular drive.

Even in your own day, Charles, you nearly fell foul of Captain

Fitzroy's belief that he could tell a man's personality from the shape of his nose, which nearly had you rejected from going on board *The Beagle*. The marvel of this is that you then went on to show to the world that individual variation is the very stuff of evolution. In the more prosaic times in which we live today, we can state categorically that there is no direct genetic link between the huge variation expressed at the individual level and the supposed genetic basis of average differences between groups, such as races.

It is technically and statistically illegitimate to extract, from what we call the 'heritability' component of a given trait, a genetic explanation for a difference in the average size of the trait between groups. The difference in averages between groups can be induced as much environmentally as genetically. Indeed, 'heritability' can be zero (meaning there is no genetic component) for a given feature when the between-group difference is genetically determined. On the other hand, heritability can be 100 per cent within a group when the between-group difference is completely environmentally determined. As I wrote some time ago, for the vast majority of traits, we could not correctly identify the group to which an individual 'belonged' after measuring some trait or other. We would be better off tossing a coin. This is because nearly ninety per cent of variation exists between individuals irrespective of the group of origin.

Human cloning: you cannot recreate yourself

Many of the same misconceptions, often espoused by professional biologists, concern the very topical issue of human cloning. Maybe this possibility will not come as a shock to you, after all that I've been relating to you about modern genetics. But there is a fear stalking the land which gets to the very heart of the misconception of the 'new genetics'. This fear is based on the erroneous belief that if we took a hundred cells out of, say, the worst dictator in our century – Hitler, the organizer of racial murder – then each cell would develop into a Hitler with the same awesome and awful behavioural traits. This prediction is a fundamental error. There is no reason to expect that precisely the same sets of genetic interactions, in exactly the same order and with precisely the same molecular partners, under the identical environmental conditions, would occur in the separate development of each and every one of our hundred 'Hitler' cells.

This argument is obvious to any person in close contact with genetically identical twins. Here all the genes are shared, not just a few, and the developmental unfolding of each identical set of genes takes place in a very common environment. Yet still the twins acquire recognizably separate personalities in all the things that matter to an individual in his or her quest for contentment. Such differentiation must be a consequence not only of the ability of multifunctional genes to interact in slightly different ways, which I've written about in earlier letters, but also of some inherent random flux in developmental circuitry about which we currently know very little. We also know very little about how minor differences in the environment influence the unique unfolding of genetic interactions in specific individuals.

There is a malleability and plasticity in shaping human form and behaviour that is a direct reflection of how genes have evolved to interact in the construction of unique individuals, generation after generation. These interactions are sensitive to minor environmental differences, making development an unpredictable science. Hitler would be more likely to produce similar behavioural phenotypes to himself if he took a hundred unrelated baby boys and subjected them to the kind of horrendous brainwashing that we know he practised.

To come back to the recent attempts at ethnic annihilation in the Balkans, which so upset me in an earlier letter, I am sure that the genetic differences between all the thugs and racists who perpetuate such horrors are no less than the differences between random samples of individuals taken from any other section of the population. Human beings can be indoctrinated, each and every one a unique phenotype, into doing the worst of evils, no matter what their genetic diversity.

At the risk of sounding completely contradictory, this is the very basis of our free will. It is our biology that makes us into genetically influenced and environmentally influenced unique individuals. This biology has emerged through all the interactive processes of modular, redundant and promiscuous genes in a constant state of long-term flux. In the same way that diversity leads to molecular tolerance and coevolution, it should also serve as a basis of tolerance in society. Deep down we are all different, which only goes to make us all the same. It is biology, through its powers to generate phenotypic uniqueness, that makes us free.

No nature, no nurture, just Chaliapin

To bring this home in a more pleasing and personally reflective manner, away from the conflicts of war, race and fundamentalism in science and religion, I want to illustrate the exercise of free will with the following example.

One of the giants of opera singing in this century was Feodor Chaliapin. As with Maria Callas, his voice had the sweep and power to express the scale of human emotions in a way beyond normal mortals. He was a Russian bass singer, plumbing the depths, but with an unexpected lightness of touch and inflection that confounded the ear. As with Callas, some thought it natural; some unnatural. Boris, Mephistopheles, The Grand Inquisitor were hewn out of rock, and the vocal bangings and scrapings of hammers and chisels as they came to life could be heard throughout.

One particular instance of vocal spontaneity in the rendition of the little-known 'Persian Love Song' by Rubinstein puts Chaliapin among the angels. I doubt if I will find the words to describe what takes place, or to convey my responses to the thread of sound that Chaliapin spins out at the very end of the song. The music throughout is flowing, gentle and heartbreaking, sung by a man who is so obviously sad, not through the loss or unattainability of love, but through the sheer inescapable and unavoidable happiness of his love. Only the great Chaliapin could have turned such a routine love song into a masterpiece of unexpected contradictions. If you are not already in tears towards the end of the song, the final eerie note takes the breath away. Here is a voice of such power and depth, yet soaring at the very last moment of indescribable emotion into a long soprano falsetto training off into the outer edges of feeling. The sound is a long, high wail of desire, highly personal, subjective and oblivious to any intended listeners. Every fibre in Chaliapin was marshalled to produce a silence without end.

I'm sure that it was not in the script: it was Chaliapin at his imperious best. Chaliapin, the uniquely, biologically formed Chaliapin, living at that moment of recording on 13 May 1931, making his spontaneous choice, a freely willed decision to switch his rich tones into an undulating falsetto, giving vent to an unearthly and terrible beauty that goes on for ever.

This decision was not written in Chaliapin's genes, in the manner of their interactions as he took shape or, I wager, in the Russian and theatrical influences that also shaped him. Chaliapin the phenotype could not be predicted in advance, except in general and uninteresting ways. The decision to let go emotionally, at that moment in time and place, was a spontaneous one-off creation of artistic expression that can only be described as a free choice. Of course, it could be decomposed into the particular neuronal firings reflecting the emotional state of Chaliapin on 13 May 1931. But such a fine-grained molecular description could not circumscribe the free spontaneity, particularity and instant beauty of Chaliapin's choice.

Chaliapin's long falsetto is neither nature nor nurture, nor any interactive component between them. It is a singularity that has no past and no future. All our individual natures, personalities, behaviours and psychologies are in the same boat. We are free to choose at each and every singular moment, if we want to. We are, on average, human, but we are, all of us, our singular selves.

Your exposed servant,

Gabriel

GLOSSARY

Adaptation An evolved feature of an organism that contributes to its viability and reproduction (shaped by natural selection).

Adoptation An evolved feature of an organism that contributes to its viability and reproduction (established by molecular drive) and that adopts some previously inaccessible component of the environment.

Allele An alternative form of a gene. For example, there can be two different forms of a given gene in a human: one from mother and one from father. Alleles differ by one or more mutations involving a change in DNA sequence.

Altruism The performance of an act that helps another related or unrelated individual, without necessarily any long-term reproductive benefit to the helper.

Base Abbreviation for nucleotide base, a building block of DNA and RNA. There are four types of base in DNA: adenine, thymine, guanine and cytosine (A, T, G and C). A, C and G also make up RNA with an additional base U for uracil.

Chromosome A structure, visible with the aid of a light microscope, in the nucleus of a cell. It consists of a long line of balls each composed of five different proteins with a continuous double helix of DNA wound around the balls from one end of the line to the other. A typical chromosome can carry enough DNA for several thousand genes and much 'junk' (non-functional) DNA. It can be replicated and each copy enters a different 'daughter' cell after the nucleus of the 'mother' cell has divided.

Concerted Evolution Many genes (whole or in part), regulatory genetic elements and stretches of junk DNA exist in multiple copies often spread on different chromosomes. If two copies are picked out at random from individuals belonging to the same species and their sequences of bases compared, they are much more similar to each other than if they had been chosen from individuals of two different species. The multiple copies of DNA are evolving in concert. Molecular drive is the process underlying the concerted evolution pattern.

Crossing-over When two chromosomes lie side by side a break can occur in each allowing one end of one chromosome to join the other end of the opposite chromosome creating two mosaic chromosomes. The break involves the two double helices of DNA. If the break occurs in precisely the same place in the two chromosomes then there is an equal crossing-over of genetic material from one chromosome to another.

Diploid The existence of two sets of chromosomes (one from each parent) and their attendant genes within a cell's nucleus.

DNA The genetic material of all life forms excluding some viruses. Consists of a double-helix of two necklaces of bases (A, T, G and C), like a twisted rope with two strands. During replication the two strands come apart and each single strand is copied to reconstitute the double-helix.

Ecology A network of interactions between living organisms and between organisms and their environment.

Enzyme Usually a protein (but sometimes RNA) that affects the rate at which a biological reaction takes place. It usually binds to and forms a temporary, intermediate complex with a substrate molecule.

Evolution The transformation of a population of organisms from one average genetic constitution to another. Called 'modification by descent' by Darwin. There are several distinct operations that can effect evolution.

Exaptation An evolved feature of an organism that is co-opted to contribute to its viability and reproduction but which did not evolve in the first instance for such a contribution. A term first coined by Stephen Jay Gould and Elizabeth Vrba.

Fitness A measure of reproductive success (number of viable and fertile offspring) of an individual (or of two sexual individuals) relative to that of other individuals.

Gene A stretch of DNA that can successfully code for a stretch of amino acids (protein) after its transcription to messenger RNA and the translation of the messenger RNA to protein.

Gene conversion The ability of a stretch of DNA (often a gene) to change the sequence of a similar stretch of DNA to its own sequence. For example, a diploid cell carrying two alleles of a gene (*A* and *a*) can produce daughter cells carrying all *A* or all *a*. Usually occurs during meiosis, and can involve copies of similar DNA whatever their distribution on the chromosomes. Sometimes the direction of conversion is biased in favour of *A* or *a*. One of the genomic mechanisms of turnover giving rise to molecular drive.

Genetic drift A change in the genetic constitution of a population due to accidents of sampling. For example, the movement of 10 individuals from one location to another location, chosen at random from a population consisting of 500 type *A* and 500 type *a*, will not necessarily transfer equal numbers of *A* and *a*. Continual fluctuations in the sizes of populations of individuals or of sperm and eggs can lead to the accidental spread of one genetic type from a mixture of types.

Gene regulation See Promoter; Regulatory Protein; Repression.

Genome The sum total of DNA (genes, regulatory elements and 'junk' DNA) in a cell.

Genotype The particular set of genetic variants (alleles) that coexist in a given individual and that partly contribute to the unique phenotypic form and behaviour of an individual.

Gonad Part of the body (for example, testis and ovary) housing the cells that undergo meiosis in the production of haploid sperm and eggs.

Genus A category of classification of organisms above the level of species. For example, *Drosophila melanogaster* is a species of fly, along with 5,000 other species, classified in the genus *Drosophila*. *Homo sapiens* is our modern human species belonging to the genus *Homo*: all other *Homo* species are extinct.

Haploid The existence of one set of chromosomes and their attendant genes within a cell's nucleus.

Homogenization The replacement of a family of genetic units by a variant member unit. There are a number of mechanisms of turnover such as gene conversion and unequal crossing-over that can lead to homogenization of repetitive genetic units. When linked to the sexual process, the variant unit may also spread through a population. The dual process of homogenization and sex gives rise to molecular drive. The unit being homogenized can be a whole gene, part of a gene, a regulatory element of a promoter, or a stretch of junk DNA.

Hybrid dysgenesis A syndrome of effects including chromosome breakage, high mutation rates, and absence of gonads in individuals of *Drosophila melanogaster* that carry the mobile P element and that do not have a genetic mechanism to stop the P elements from jumping from one chromosome to another chromosome.

Junk DNA Most genomes of species above the bacteria consist of vast excesses of DNA that do not make sense *vis-à-vis* the proper readout of the genetic code required to make sensible proteins. Most excess DNA is repetitive at one level or another due to a variety of non-Mendelian DNA turnover mechanisms, revealing homogenized families of repeats (concerted evolution). Much of the excess DNA has no known function. It can exceed the gene requirements of a species by several orders of magnitude. For example, in humans 95–99 per cent of the genome is considered to be 'junk'. The genes and their promoters are small islands in a vast ocean of scrambled sequences of junk DNA. The evolutionary history of junk DNA is predominantly influenced by the process of molecular drive and neutral genetic drift rather than natural selection.

KP element A short P element that has lost most of its middle section containing the gene for the jumping (transposase) enzyme. KP elements can repress hybrid dysgenesis probably through their direct interference with the jumping mechanism of full-size P elements.

Ligase An enzyme capable of joining up two free ends of DNA to make one continuous double helix.

Macromolecule Very large molecules such as DNA, RNA, proteins and fats which consist of strings of smaller molecules that are the basic building blocks.

Meiosis A process of cell division of a diploid 'mother' cell to produce haploid 'daughter' cells. For example, haploid sperm or eggs, each containing one set of chromosomes, are produced from diploid cells containing two sets of chromosomes, one set from each parent. Meiosis ensures that each and every sperm or egg contains a mixture of the original maternal and paternal chromosomes, while ensuring that one copy of each type of chromosome is in each sperm or egg. For example, the human diploid cell contains two sets of 23 different types of chromosome and after meiosis each haploid sperm or egg contains one of each type. However, because of the random shuffling of chromosomes during meiosis it is highly improbable that a complete set of 23 derived from either the individual's mother or father will enter the same sperm or egg.

Mendelian segregation The observation that alleles of a gene segregate to the next generation according to the laws of chance. This is because of the random assortment of chromosomes during the formation of sperm and eggs by meiosis and the random fusion of haploid sperm and eggs to reconstitute the next set of diploid individuals. Assumes that the segregation of a gene and that of the chromosome on which it resides are in phase. However, only rare genes, not subject to one or other non-Mendelian mechanism of DNA turnover, will segregate along strict Mendelian lines. The slow operations of turnover mean that Mendelian segregation is, nevertheless, observed in most genes when examined in small numbers of progeny or over a small number of generations, hence not allowing sufficient numbers of progeny or time to elapse for the effects of non-Mendelian segregation to be measurable.

Microbe A loose term, generally encompassing unicellular bacteria in which there is no well-defined nucleus and no complex chromosome structure as found in organisms with true nuclei. Meiosis and true sex do not occur in microbes although microbes can partially exchange genetic material by other means.

Mitosis A process of cell division by which a cell divides into two ensuring that each 'daughter' cell receives the full complement of chromosomes, either haploid or diploid, that exists in the 'mother' cell. All chromosomes are replicated before the cell and its nucleus divide into two.

Modularity The construction of complex structures through the use of repetitive modular parts, as in the game of Lego. Modules can exist at several biological levels such as DNA (genes, part of genes and promoters), proteins, genetic operations during development, cells and organs. For example, the heart, brain and backbones of vertebrates are composed of modules. Modules are functionally autonomous units that can combine with each other in a variety of permutations either freely or through co-option. Modules of genes, promoters and proteins are often shared by unrelated genes, promoters and proteins indicating their ability to move from one genetic region to another via a variety of turnover mechanisms.

Molecular coevolution Molecular coevolution is the process behind the observation that two interacting molecules A and B function successfully in species Y and their variant forms A' and B' do likewise in species X. However A cannot successfully function with B' and B cannot successfully function with A'. All biological functions require molecules to interact with each other. The interaction depends on the molecular shapes of the molecules which in turn depend on their particular chemical make-up. In the case of DNA and proteins this depends on their base or amino-acid composition. A mutation in one of a pair of interacting molecules can disturb the appropriate contact required for function. However, if a second mutation occurs in the second partner that is capable of compensating for the mutation in the first partner, then contact and function are maintained. The two interacting molecules have coevolved. Molecular coevolution occurs frequently when one partner exists in multiple copies, for example, the multiple DNA binding sites that constitute a given gene's promoter. Such redundancy permits one or other mechanism of turnover to operate leading ultimately to the spread of the reconstructed promoter through a sexual population (molecular drive). The protein module that binds to promoter binding-sites (for example the homeodomain) might need to mutate in order to compensate for the spread of mutant binding sites. The role of natural selection would be to promote any available allele of the gene coding for the homeodomain protein that binds successfully and appropriately with the new set of binding sites.

Molecular drive A process capable of changing the average genetic composition of a sexual population through the generations as a consequence of a number of non-Mendelian mechanisms of DNA turnover (for example, transposition, gene conversion, unequal crossing-over and others). Such mechanisms increase or decrease the number of copies of new or variant genetic elements in an individual notwithstanding the often wide distribution of elements among the chromosomes. So if two copies of a genetic element are made in an individual that started with one copy and if the two copies reside on different chromosomes, then the subsequent shuffling of chromosomes by sex ensures that each copy enters a different individual at the next generation in each of which more copies of the element can be made. The combination of non-mendelian DNA turnover in multigene families, regulatory genetic elements, internally repetitious genes or junk DNA, when coupled to sex, ensures the spread (molecular drive) through a population of novel genetic variants over evolutionary time. Any inherent bias in a turnover mechanism speeds up the rate of spread. Molecular drive is operationally distinct from natural selection and genetic drift. The dynamics of molecular drive at the population level facilitates molecular coevolution and adoptation for all biological functions exhibiting TRAM features, (turnover, redundancy and modularity).

Mutation A change in the chemical composition of a genetic element. For example, substitutions of one or more nucleotide bases with alternative bases; gains or losses of bases; switching the orientation of bases; and moving genetic elements from one genomic position to another.

Natural selection A means for changing the average genetic composition of a population over time, as a consequence of differences in reproductive success of individual phenotypes and the inheritance of the genes partly responsible for such differences by the next generation of phenotypes. Gives rise to adaptations. Natural selection is operationally distinct from molecular drive and genetic drift.

Neutral drift See genetic drift. For genetic drift to operate independently of natural selection the variant genetic element needs to be neutral with respect to its effect on the reproductive success of individuals carrying the variant genetic element.

Non-Mendelian The increase or decrease of genetic units in the lifetime of an individual as a consequence of a variety of mechanisms of DNA turnover. The long-term segregation of genetic elements that do not obey Mendel's rules.

Nucleotide See Base.

P element A mobile genetic element in species of *Drosophila* that induces the phenomenon of hybrid dysgenesis.

Phenotype All aspects of an organism over and above the genetic material. For example, proteins, developmental and metabolic processes, form and behaviour. Uniquely shaped for each individual by complex networks of genetic interactions that are also intimately affected by local environmental conditions. The phenotype can reproduce autonomously, asexually or sexually.

Promoters A region of DNA, most often near the beginning of a gene, that contains many short stretches of bases that bind to regulatory proteins. The combination and number of bound proteins determine the time, place and degree of transcription of the gene. The binding sites can differ in type for different forms of regulatory proteins and there are often multiple copies of each type. The extensive sharing of binding sites between unrelated promoters reveals the modular nature of the sites and their ability to move around the genetic material by one or other mechanisms of DNA turnover.

Protein A string of amino acids that can fold into a three-dimensional shape capable of performing a given biological function such as an enzyme or regulatory protein. The types and order of amino acids in a given protein is a direct outcome, via the genetic code, of the types and order of nucleotide bases in a given gene.

Redundancy Means both no longer functionally required and existing in multiple copies. The meanings are bound up with each other in that the more extra copies of an element there are, the less requirement there is for any given element. Redundancy exists at all levels in the genetic material: multiple copies of short binding sites in gene promoters; of stretches of DNA within genes; of whole genes; of complete clusters of genes; and whole chromosomes or sets of chromosomes with all their attendant genes. At the level of the phenotype redundancy in proteins is a direct consequence of redundancy at the gene level; and in developmental operations in which two different processes can achieve the same end. Redundant genetic systems can be buffered from the effects of a mutation in a repetitive unit. The extent of buffering depends on the total number of copies and their functional necessity.

Regulatory protein A protein that contains modules (stretches of amino-acids) capable of binding to short stretches of DNA (binding sites) in the promoters of genes or to other proteins or RNA molecules involved with gene regulation.

Repression Genetic systems can be turned 'off' (repressed) as well as turned 'on' (activated). For example P elements are prevented from jumping by a variety of genetic processes. Gene transcription can be repressed by a specific regulatory protein binding to particular sites in a promoter.

Ribosome A complex structure consisting of over 100 proteins and several types of RNA that is intimately involved in translating the base sequence of messenger RNA to the amino-acid sequence of a protein.

RNA There are several types of RNA. For example, messenger RNA acts as an intermediate between DNA and protein during translation. Ribosomal RNA, coded by ribosomal RNA genes (rDNA), is structurally and functionally required in the ribosome where translation takes place. See also DNA.

Segregation The distribution of genes and chromosomes among individuals from one generation to the next.

Sexual reproduction An alternating haploid–diploid cycle. Sex consists of two parts: the reduction of diploid cells (two sets of chromosomes) to haploid cells (one set of chromosomes) in the formation of sperm or egg by meiosis and the reconstitution of the diploid state by fusion of an egg with a sperm.

Slippage A mechanism of DNA turnover involving the two strands of nucleotides that comprise the double helix. A tandem array of short repetitive sequences (each repeat usually less than 10 bases) can cause the two strands to slip such that only 9 repeats of each strand are matched together leaving one repeat of each unmatched. The unmatched repeat can be either deleted by special enzymes or used as a template to generate a complementary matching repeat on the opposite strand. Slippage is a continual process of gain and loss of repeats. As such it can induce the homogenization of an array by any given variant repeat. See Crossing-over. Slippage is the most frequently occurring mechanism of turnover in genomes, often within the body of a gene generating repeats that can be of different types and interspersed one with another.

Species A population containing individuals that are sexually compatible but that are usually not compatible with individuals of another species. Species in clonal or asexual organisms are difficult to characterise and define.

Substrate A molecule to which an enzyme binds in order to bring about a biological reaction.

TRAM Genetic systems that have the features of non-Mendelian Turnover, copy number and functional Redundancy And Modularity. To date all regulatory regions (promoters) and genes, that have been examined in detail at the molecular level, have TRAM characteristics. As such, part of their evolutionary history will have been influenced by the molecular drive process.

Transcription The first step in the reading of the genetic code from gene to protein. Involves the formation of a single strand of messenger RNA from the double-stranded helix of DNA. The sequence of bases in the messenger RNA complements the equivalent sequence of one of the DNA strands. Transcription (readout) of a gene begins in a given place and time and with a given efficiency depending on the binding of regulatory proteins to the gene's promoter.

Translation The second step in reading the genetic code. This involves translating a sequence of bases of the messenger RNA to a sequence of amino acids of a protein. A given triplet of bases codes for a given amino acid. Some amino acids are coded by more than one triplet. Translation occurs in the ribosome, involving many different proteins and the ribosomal RNA of the rDNA genes.

Transposase An enzyme capable of making a transposable element, such as the P element, jump from one genomic position to another, in conjunction with other enzymes involved with DNA metabolism. The gene for transposase can be contained within the mobile element itself.

Turnover Mechanisms of DNA rearrangement leading to the continual gain and loss of genetic material. For example, gene conversion, unequal crossing-over, slippage, transposition and others. Mechanisms generate non-Mendelian patterns of segregation and can promote or demote the frequency of genetic variants in a sexual population. Mechanisms can often occur simultaneously in a given region of DNA as one mechanism (for example, slippage) can trigger off another (for example, gene conversion). There are few, if any, regions of the genetic material, including genes and their promoters, that are not subject to one or other mechanism of turnover. Rates of turnover vary widely but generally lie between the basic rate of mutation and the rate at which sex randomizes chromosomes between generations. Such rate differentials have implications for the gradual and cohesive ways by which turnover affects the genetic composition of a population. (See also Molecular drive and Homogenization).

FURTHER READING

Books

Akam, M., Holland, P., Ingham, P. & Wray, G. (eds.) *The Evolution of Developmental Mechanisms* (*Development* supplement, Company of Biologists, 1994).

Bateson, William M. *Materials for the Study of Variation* (Macmillan, 1894).

Bateson, Patrick & Martin, Paul *Design for a Life: How Behaviour Develops* (Cape, 1999).

Coen, Enrico *The Art of Genes* (Oxford University Press, 1999).

Darwin, Charles *On the Origin of Species* (Murray, 1859).

Dover, G. A. & Flavell, R. B. (eds.) *Genome Evolution* (Academic Press, 1982).

Eldredge, N. *Reinventing Darwin: The Great Debate at the High Table of Evolutionary Theory* (Wiley, 1995).

Fedoroff, N. & Botstein, D. (eds.) *The Dynamic Genome: Barbara McClintock's Ideas in the Century of Genetics* (Cold Spring Harbor Laboratory Press, 1992).

Fisher, Ronald A. *The Genetical Theory of Natural Selection* (Clarendon Press, 1930).

Ghysen, A. (ed.) Developmental Genetics in *Drosophila*: Special Issue Dedicated to Antonio Garcia-Bellido, *Int. J. Devel. Biol.* **42,** 539 (1998).

Gould, Stephen Jay *Wonderful Life* (Hutchinson Radius, 1989).

Gould, Stephen Jay *The Mismeasure of Man* (Norton, 1981).

Hoelzel, A. R. & Dover, G. A. *Molecular Genetic Ecology* (Oxford University Press, 1991).

Hull, D. L. & Ruse, M. (eds.) *The Philosophy of Biology* (Oxford University Press, 1998).

Huxley, Julian *Evolution: The Modern Synthesis* (Allen & Unwin, 1942).

Jackson, M. S., Strachan, T. & Dover, G. A. (eds.) *Human Genome Evolution* (Bios Scientific, 1996).

John, B. & Miklos, C. *The Eukaryotic Genome in Development and Evolution* (Allen & Unwin, 1988).

Jones, S. *Almost Like a Whale: The* Origin of Species *Updated* (Doubleday, 1999).

Kimura, Motoo *The Neutral Theory of Evolution* (Cambridge University Press, 1983).

Kohn, Marek *As We Know It: Coming to Terms With an Evolved Mind* (Granta, 1999).

Lawrence, P. A. *The Making of a Fly: The Genetics of Animal Design* (Blackwell Science, 1992).

Levins, Richard & Lewontin, Richard *The Dialectical Biologist* (Harvard University Press, 1985).

Lewontin, R. C., Rose, S. & Kamin, L. J. *Not in Our Genes* (Pantheon, 1984).

Lloyd, Elisabeth *The Structure and Confirmation of Evolutionary Theory* (Princeton University Press, 1994).

Mayr, Ernst *Evolution and the Diversity of Life* (Harvard University Press, 1976).

Monod, Jacob *Chance and Necessity* (Knopf, 1971).

Ohno, Susumo *Evolution by Gene Duplication* (Springer, 1970).

Ohta, T. *Evolution and Variation of Multigene Families* (Springer, 1980).

Oyama, Susan *The Ontogeny of Information* (Cambridge University Press, 1985).

Pollack, R. *Signs of life: The Language and Meanings of DNA* (Penguin, 1994).

Raff, Rudolf, A. *The Shape of Life* (Chicago University Press, 1996).

Ridley, Matt *Genome: An Autobiography of a Species in 23 Chapters* (Fourth Estate, 1999).

Rose, Steven, *Lifelines* (Penguin, 1999).

Rose, Steven & Rose, Hilary (eds.) *Alas, Poor Darwin: The Case Against Evolutionary Psychology* (Random House, in press).

Schlichting, Carl & Pigliucci, Massimo *Phenotypic Evolution* (Sinauer, 1998).

Sherratt, D. J. (ed.) *Mobile Genetic Elements* (Oxford University Press, 1985).

Smolin, Lee *The Life of the Cosmos* (Weidenfeld & Nicolson, 1997).

Sober, Elliott & Wilson, David Sloan *Unto Others: The Evolution and Psychology of Unselfish Behaviour* (Harvard University Press, 1998).

Wilson, Edward O. *On Human Nature* (Penguin, 1978).

Selected Publications

Akam, M. Hox genes: From master genes to micromanagers. *Current Biology* **8**, R676–R678; 1998.

Akam, M. The molecular basis for metameric pattern in the *Drosophila* embryo. *Development* **101**, 1–22; 1987.

Akam, M. Hox and HOM: Homologous gene clusters in insects and vertebrates. *Cell* **57**, 347–349; 1989.

Aparicio, S. Exploding vertebrate genomes. *Nature Genetics* **18**, 301–303; 1998.

Aparicio, S. *et. al.* Organization of the *Fugu rubripes Hox* clusters: Evidence for continuing evolution of vertebrate *Hox* complexes. *Nature Genetics* **16**, 79–83; 1997.

Arnheim, N. Concerted evolution of multigene families in *Evolution of Genes and Proteins* (ed. M. Nei & R.K. Koehn) 38–61 (Sinauer, 1983).

Arnheiter, H. Eyes viewed from the skin. *Nature* **391**, 632–633; 1998.

Arnone, M. I. & Davidson, E. H. The hardwiring of development: Organization and function of genomic regulatory systems. *Development* **124**, 1851–1864; 1997.

Averof, M. Same *Hox* genes, different body plans. *Current Biology* **7**, R634–R636; 1997.

Bachiller, D. Macias, A., Duboule, D. & Morata, G. Conservation of a functional hierarchy between mammalian and insect *Hox/HOM* genes. *EMBO J.* **13:8**, 1930–1941; 1994.

Baltimore, D. Gene conversion: some implications for immunoglobulin genes. *Cell* **24**, 592; 1981.

Bateson, P. P. G. The active role of behaviour in evolution in *Evolutionary Processes and Metaphors* (ed. M. W. Ho & S.W. Fox) 192–207 (Wiley, 1998).

Bork, P. Mobile modules and motifs. *Current Opinion in Structural Biology* **2**, 413–421; 1992.

Brakefield, P. M. & French, V. Butterfly wings: The evolution of development of colour patterns. *BioEssays* **21**, 391–401; 1999.

Brenner, S., Dove, W., Herskowitz, I. & Thomas, R. Genes and development: Molecular and logical themes. *Genetics* **126**, 479–786; 1990.

Britten, R. J. Underlying assumptions of developmental models. *Proc. Natl. Acad. Sci. USA* **95**, 9372–9377; 1998.

Carroll, S. B. Homeotic genes and the evolution of arthropods and chordates. *Nature* **376**, 479–485; 1995.

Coates, M. I. & Cohn, M. J. Fins, limbs, and tails: outgrowths and axial patterning in vertebrate evolution. *BioEssays* **20**, 371–381; 1998.

Coen, E. S. Floral symmetry. *EMBO Journal* **15:24**, 6777–6788; 1996.

Cooke, J., Nowak, M. A., Boerlijst, M. & Maynard-Smith, J. Evolutionary origins and maintenance of redundant gene expression during metazoan development. *Trends in Genetics* **13:9**, 360–364; 1997.

Coyne, J. A. & Orr, H. A. The evolutionary genetics of speciation. *Phil. Trans. Roy. Soc. Lond.* **B353**, 287–305; 1998.

Dahl, E., Koseki, H. & Balling, R. *Pax* genes and organogenesis. *BioEssays* **19:9**, 755–765; 1997.

Dahmann, C. & Basler, K. Compartment boundaries: At the edge of development. *Trends in Genetics* **15**, 320–326; 1999.

Davidson, E. H. Molecular biology of embryonic development: How far have we come in the last ten years? *BioEssays* **16**, 9603–9615; 1994.

Davidson, E. H., Peterson, K. J. & Cameron, R. A. Origin of bilaterian body plans: Evolution of developmental regulatory mechanisms. *Science* **270**, 1319–1325; 1995.

Dawkins, R. Parasites, desiderata lists and the paradox of the organism. *Parasitology* **100**, S63–S73; 1990.

Dickinson, W. J. The evolution of regulatory genes and patterns in *Drosophila*. *Evolutionary Biology* **25**, 127–173; 1990.

Dover, G. A. Molecular drive: A cohesive mode of species evolution. *Nature* **299**, 111–117 (1982).

Dover, G. A. & Flavell, R. B. Molecular coevolution: DNA divergence and the maintenance of function. *Cell* **38**, 623–624; 1984.

Dover, G. A. Molecular drive in multigene families: How biological novelties arise, spread and are assimilated. *Trends in Genetics* **2**, 159–165; 1986.

Dover, G. A. DNA turnover and the molecular clock. *J. Mol. Evol.* **26**, 47–58; 1987.

Dover, G. A. Evolving the improbable. *Trends in Ecol. Evol.* **3**, 81–84; 1988.

Dover, G. A. Observing development through evolutionary eyes: A practical approach to molecular coevolution. *Bioessays* (Special Issue: Evolution and Development) **14**, 281–287; 1992.

Dover, G. A., Ruiz Linares, A., Bowen, T. & Hancock, J. M. The detection and quantification of concerted evolution and molecular drive. *Methods in Enzymology* **224**, 525–541; 1993.

Dover, G. A. The evolution of genetic redundancy for advanced players. *Current Opinion in Genetics and Development* **3**, 902–910; 1993.

Dover, G. A. Human evolution: our turbulent genes and why we are not chimps. In *The Human Inheritance: Genes, Languages and Evolution* (ed. B. Sykes) (Oxford University Press, in press).

Dover, G. A. Anti-Dawkins. In *Alas, Poor Darwin: The Case Against Evolutionary Psychology* (eds. S. Rose & H. Rose) (Random House, in press).

Duboule, D. & Wilkins, A. The evolution of bricolage. *Trends in Genetics* **14**, 254–59; 1998.

Edelman, G. M. & Gally, J. A. Antibody structure, diversity and specificity. *Brookhaven Symp. Biol* **21**, 328–343; 1968.

Elder, J. F. & Turner, B. J. Concerted evolution of repetitive DNA sequences in eukaryotes. *Quarterly Review of Biology* **70:3**, 297–320; 1995.

Engels, W. R. The origins of P elements in *Drosophila melanogaster*. *BioEssays* **14**, 10681–10686; 1992.

Fedoroff, N. V. About maize transposable elements and development. *Cell* **56**, 181–191; 1989.

Fryxell, K. J. The coevolution of gene family trees. *Trends in Genetics* **12:9**, 364–369; 1996.

Gally, J. A. & Edelman, G. M. The genetic control of immunoglobulin synthesis. *Ann. Rev. Genetics* **6**, 1–46; 1972.

Garcia-Bellido, A. Genetic control of wing disc development in *Drosophila*. *Ciba Foundation Symposium* **29**, 161–182; 1975.

Garcia-Bellido, A. Symmetries throughout organic evolution. *Proc. Natl. Acad. Sci. USA* **93**, 14229–14232; 1996.

Garcia-Bellido, A. The idea of progress in *Progress in Biological Evolution* (ed. W. de Gruyter) 175–199 (Berlin, 1997).

Garcia-Bellido, A., Lawrence, P. A. & Morata, G. Compartments in animal development. *Scientific American* **241**, 102–110; 1979.

Garcia-Bellido, A. C. & Garcia-Bellido, A. Cell proliferation in the attainment of constant sizes and shapes: The entelechia model. *Int. J. Dev. Biol.* **42**, 353–362; 1998.

Gehring, W. J. Homeoboxes in the study of development. *Science* **236**, 1245–1252; 1987.

Gehring, W. J. & Ikeo, K. *Pax6*: Mastering eye morphogenesis and eye evolution. *Trends in Genetics* **15**, 371–377; 1999.

Gellon, G. & McGinnis, W. Shaping animal body plans in development and evolution by modulation of *Hox* expression patterns. *BioEssays* **20**, 116–125; 1998.

Gibson, G. Insect evolution: Redesigning the fruitfly. *Current Biology* **9**, R86–R89; 1999.

Gould, S. J. The exaptive excellence of spandrels as a term and prototype. *Proc. Natl. Acad. Sci. USA* **94**, 10750–10755; 1997.

Gould, S. J. & Eldredge, N. Punctuated equilibrium comes of age. *Nature* **18**, 223–227: 1993.

Gould, S. J. & Lewontin, R. C. The spandrels of San Marco and the Panglossian paradigm: A critique of the adaptationist programme. *Proc. R. Soc. Lond.* **B205**, 581–598; 1979.

Gould, S. J. & Vrba, E. S. Exaptation: A missing term in the science of form. *Paleobiology* **8**, 4–15; 1982.

Gray, S. & Levine, M. Transcriptional repression in development. *Current Opinion in Cell Biology* **8**, 358–364; 1996.

Hancock, J. M. The contribution of slippagelike processes to genome evolution. *J. Mol. Evol.* **41**, 1038–1047; 1995.

Hancock, J. M. Simple sequences and the expanding genome. *BioEssays* **18:5**, 421–425; 1996.

Holland, P. W. H. Homeobox genes and segmentation: Co-option, co-evolution, and convergence. *Developmental Biology* **1**, 1–10; 1990.

Hood, L. J., Campbell, J. H. & Elgin, S. C. R. The organisation, expression, and evolution of antibody genes and other multigene families. *Ann. Rev. Genet.* **9**, 305; 1975.

Hurst, L. D. The evolution of genomic anatomy. *Trends in Ecol. Evol.* **4**, 3108–3112; 1999.

Jacob, F. Evolution and tinkering. *Science* **196**, 1161–1166; 1977.

Jeffreys, A . J., Wilson, V. & Thein, S. L. Hypervariable minisatellite regions in human DNA. *Nature* **314**, 67–73; 1985.

Jeffreys, A. J. et al. Minisatellite repeat coding as a digital approach to DNA typing. *Nature* **354**, 204–209; 1991.

Jeffreys. J. et al. Complex gene conversion events in germline mutation at human minisatellites. *Nature Genetics* **6**, 136–145; 1994.

Jockusch, E. L. & Nagy, L. M. Insect evolution: How did insect wings originate? *Current Biology* **7**, R358–R361; 1997.

Kashi, Y., King, S. & Soller, M. Simple sequence repeats as a source of quantitative genetic variation. *Trends in Genetics* **13:2**, 74–78; 1997.

Kazazian, H. H. Mobile elements and disease. *Current Opinion in Genetics and Development* **8**, 343–350; 1998.

Kenyon, C. If birds can fly, why can't we? Homeotic genes and evolution. *Cell* **78**, 175–180; 1994.

Keys, D. N. et al. Recruitment of a *hedgehog* regulatory circuit in butterfly eyespot evolution. *Science* **283**, 532–534; 1999.

Kidwell, M. G. The evolutionary history of the *P* family of transposable elements. *Journal of Heredity* **85**, 339–346; 1994.

Kirschner, M. & Gerhart, J. Evolvability. *Proc. Natl. Acad. Sci. USA* **95**, 8420–8427; 1998.

Krumlauf, R. *Hox* genes in vertebrate development. *Cell* **78**, 191–201; 1994.

Kyriacou, C. P. & Hall, J. C. Interspecific genetic control of courtship song production and reception in *Drosophila*. *Science* **232**, 494–497; 1986.

Lamb, B. C. Gene conversion disparity in yeast: its extent, multiple origins, and effects on allele frequencies. *Heredity* **80**, 538–552; 1996.

Lamb, B. C. & Helmi, S. The extent to which gene conversion can change allelic frequencies in populations. *Genet. Res.* **39**, 199; 1982.

Lawrence, P. A. Homeobox genes: Their function in *Drosophila* segmentation and pattern formation. *Cell* **78**, 181–189; 1994.

Leibler, S. & Barkai, N. Robustness in simple biochemical networks. *Nature* **387**, 913–917 (1997).

Lewin, R. Why is development so illogical? *Science* **224**, 1327–1329; 1984.

Li, X. & Noll, M. Evolution of distinct developmental functions of three *Drosophila* genes by acquisition of different *cis*-regulatory regions. *Nature* **367**, 83–87; 1994.

Lowe, C. J. & Wray, G. A. Radical alterations in the roles of homeobox genes during enchinoderm evolution. *Nature* **389**, 718–721; 1997.

Mann, R. S. & Affolter, M. *Hox* proteins meet more partners. *Current Opinion in Genetics and Development* **8**, 423–429; 1998.

Martinez-Arias, A. On the developmental and evolutionary role of some genes from the *ant-c* complex in *Molecular Approaches to Developmental Biology* (eds. R. A. Firtel & E. H. Davidson) 131–145 (A. R. Liss, 1987).

Martinez-Arias, A. *Wnt* Signalling: Pathway or network? *Current Opinion in Genetics and Development* **9**, 447–454; 1999.

McAdams, H. H. & Arkin, A. Stochastic mechanisms in gene expression. *Proc. Natl. Acad. Sci. USA* **94**, 814–819; 1997.

Metzenberg, S., Joblet, C., Verspieren, P. & Agabian, N. Ribosomal protein L25 from *Trypanosoma brucei*: Phylogeny and molecular co-evolution of an rRNA-binding protein and its rRNA binding site. *Nucleic Acids Research* **21**, 4936–4940; 1993.

Miklos, G. L. G. Molecules and cognition: The latterday lessons of levels, language and lac. *Journal of Neurobiology* **24:6**, 842–890; 1993.

Miklos, G. L. G. & Rubin, G. M. The role of the genome project in determining gene function: insights from model organisms. *Cell* **86**, 521–529; 1996.

Missler, M. & Sudhof, T. C. Neurexins: Three genes and 1001 products. *Trends in Genetics* **14:1**, 20–25; 1998.

Murray, J. A. H., Cesareni, G. & Argos, P. Unexpected divergence and molecular coevolution in yeast plasmids. *J. Mol. Biol.* **200**, 1–7; 1988.

Nagylaki, T. & Petes, T. D. Intrachromosomal gene conversion and the maintenance of sequence homogeneity among repeated genes. *Genetics* **100**, 315; 1982.

Nino, J. The evolutionary design of error-rates, and the fast fixation enigma. *Origins of Life and Evolution of the Biosphere* **27**, 609–621; 1997.

Niswander, L. Legs to wings and back again. *Nature* **398**, 751–752; 1999.

Nowak, M. A., Boerlijst, M. C., Cooke, J. & Maynard Smith, J. Evolution of genetic redundancy. *Nature* **388**, 167–171; 1997.

Palopoli, M. F. & Patel, N. H. Evolution of interaction between *Hox* genes and a downstream target. *Current Biology* **8**, 587–590; 1998.

Parsch, J., Tanda, S. & Stephan, W. Site-directed mutations reveal long-range compensatory interactions in the *Adh* gene of *Drosophila melanogaster. Proc. Natl. Acad. Sci. USA* **94**, 928–933; 1997.

Patel, N. H. Developmental evolution: Insights from studies of insect segmentation. *Science* **266**, 581–589; 1994.

Pawson, T. Protein modules and signalling networks. *Nature* **373**, 573–550; 1995.

Pearson, C. E. & Sinden, R. R. Trinucleotide repeat DNA structures: Dynamic mutations from dynamic DNA. *Current Opinion in Structural Biology* **8**, 321–330; 1998.

Peixoto, A. A. *et al.* Molecular coevolution within a *Drosophila* clock gene. *Proc. Natl. Acad. Sci. USA* **95**, 4475–4480; 1998.

Percival-Smith, A., Muller, A., Affolter, M. & Gehring, W. J. The interaction with DNA of wild-type and mutant *fushi-tarazu* homeodomains. *EMBO J.* **9**, 3967–3974; 1990.

Purugganan, M. D. The molecular evolution of development. *BioEssays* **20**, 700–711; 1998.

Richardson, M. K. Vertebrate evolution: The developmental origins of adult variation. *BioEssays* **21**, 604–613; 1999.

Rosato, E. *et al.* Mutational mechanisms, phylogeny, and evolution of a repetitive region within a clock gene of *Drosophila melanogaster. J. Mol. Evol.* **42**, 392–408; 1996.

Schaffner, G. *et al.* Redundancy of information in enhancers as a principle of mammalian transcription control. *J. Mol. Biol.* **201**, 81–90; 1988.

Scott, M. P. Intimations of a creature. *Cell* **79**, 1121–1124; 1994.

Sharkey, M., Graba, Y. & Scott, M. P. *Hox* genes in evolution: protein surfaces and paralog groups. *Trends in Genetics* **13:4**, 145–151; 1997.

Smith, G. P. Unequal crossover and the evolution of multigene families. *Cold Spring Harbor Lab. Symp. Quant. Biol.* **38**, 507; 1973.

Sommer, R. J., Tautz, D. & Tautz, D. Evolution of segmentation genes in insects. *Trends in Genetics* **11**, 23–27; 1995.

Tautz, D. Selector genes, polymorphisms, and evolution. *Science* **271**, 160–161; 1996.

Wagner, A. The fate of duplicated genes: Loss or new function? *BioEssays* **20**, 785–788; 1998.

Wagner, G. P. Homologues, natural kinds and the evolution of modularity. *Am. Zool.* **36**, 36–43; 1996.

Wake, D. B., Roth, G. & Wake, M. On the problem of stasis in organismal evolution. *J. Theor. Biol.* **101**, 211–224; 1983.

Warren, R. & Carroll, S. Homeotic genes and diversification of the insect body plan. *Current Opinion in Genetics and Development* **5**, 459–465; 1995.

Weatherbee, S. C., Nijhout, H. F., Grunert, L. W. & Halder, G. *Ultrabithorax* function in butterfly wings and evolution of insect wing patterns. *Current Biology* **9**, 109–115; 1999.

Weintraub, H. The MyoD family and myogenesis: Redundancy, networks, and thresholds. *Cell* **75**, 1241–1244; 1993.

Wheeler, D. A. *et al.* Molecular transfer of a species-specific behaviour from
    Drosophila simulans to *Drosophila melanogaster*. *Science* **251**, 1082–1085; 1991.

Wilson, A. C. Gene regulation in evolution. In *Molecular Evolution* (ed. F. J. Ayala)
    225–334 (Sinauer, 1976).

Wright, S. The shifting balance theory and macroevolution. *Ann. Rev. Genet.* **16**,
    1–19; 1982.

Zuker, C. S. On the evolution of eyes: Would you like it simple or compound?
    Science **265**, 742–788; 1994.

INDEX